Animal Intimacies

Animal Lives

Jane C. Desmond, Directing Editor; Barbara J. King and Kim Marra, Series Editors

Books in the series

Displaying Death and Animating Life: Human-Animal Relations in Art, Science, and Everyday Life
by Jane C. Desmond

Voracious Science and Vulnerable Animals: A Primate Scientist's Ethical Journey
by John P. Gluck

The Great Cat and Dog Massacre: The Real Story of World War Two's Unknown Tragedy
by Hilda Kean

Animal Intimacies

Interspecies Relatedness in India's Central Himalayas

RADHIKA GOVINDRAJAN

The University of Chicago Press
Chicago and London

The University of Chicago Press, Chicago 60637
The University of Chicago Press, Ltd., London
© 2018 by The University of Chicago
Published 2018
Printed in the United States of America

27 26 25 24 23 22 21 20 19 18   1 2 3 4 5

ISBN-13: 978-0-226-55984-1 (cloth)
ISBN-13: 978-0-226-55998-8 (paper)
ISBN-13: 978-0-226-56004-5 (e-book)
DOI: https://doi.org/10.7208/chicago/9780226560045.001.0001

Library of Congress Cataloging-in-Pubication Data

Names: Govindrajan, Radhika, author.
Title: Animal intimacies : interspecies relatedness in India's Central Himalayas /
    Radhika Govindrajan.
Other titles: Animal lives (University of Chicago Press)
Description: Chicago : The University of Chicago Press, 2018. | Series: Animal lives
Identifiers: LCCN 2017049887 | ISBN 9780226559841 (cloth : alk. paper) | ISBN
    9780226559988 (pbk. : alk. paper) | ISBN 9780226560045 (e-book)
Subjects: LCSH: Human-animal relationships—India—Uttarakhand. | Wildlife
    conservation—India—Uttarakhand. | Animals—India—Uttarakhand—Religious
    aspects.
Classification: LCC QL85 .G68 2018 | DDC 333.95/416095451—dc23
LC record available at https://lccn.loc.gov/2017049887

♾ This paper meets the requirements of ANSI/NISO Z39.48-1992
(Permanence of Paper).

Contents

Acknowledgments

In writing this book about the nature and forms of relatedness, I have been constantly reminded of how grateful I am for the numerous kith and kin whose immense generosity, support, and love have kept me going in this endeavor. My first debt of gratitude is to all the people in Kumaon who folded me into their lives and worlds and never once allowed me to feel lonely. *Ijja*, *babu*, and all the *chachis* and *chachas*, *mamis* and *mamas*, *didis*, *dadas*, and *behnis*, *mausis* and *mausas*, *tais* and *taus*, *buas* and *phuphas*, *aamas* and *bubus*, *bhabhis*, and numerous *dagdees* who are now such important nodes in my network of kin have taught me what it means to live in relation to others. Thank you for showing me how to bear the weight of attachments with joy, care, and thought. This research would simply not have been possible without the constant friendship and support of Girish (MD). The deep curiosity and joy with which he approaches the world around him is infectious, and I owe him special thanks for bringing the same joy and curiosity to my work.

This book draws on research that I first conducted while studying at Yale University. More than anyone else, my teacher and advisor, K. Sivaramakrishnan, has taught me the value of melding intellectual rigor and generosity with humane and supportive mentorship. I hope he can see that his brilliant and sage counsel shapes each page that follows. Helen Siu urged me to draw on my early training as a historian and constantly ground my ethnographic reflections in historical study; I hope I have done justice to the example she set. James Scott's exceptional and seemingly boundless intellectual curiosity pushed me into new terrain that I might not have tread if it had not been for his unwavering support for and excitement about this project. I was fortunate enough to be part of a community of inspirational teachers and scholars at Yale who have shaped my thoughts and interests in profound ways.

Classes and conversations with Jafari Allen, Bernard Bate, Sean Brotherton, Kamari Clarke, Michael Dove, Narges Erami, Joseph Errington, Erik Harms, Robert Harms, Karen Hébert, Marcia Inhorn, William Kelly, Michael Mc-Govern, Alan Mikhail, Karen Nakamura, Douglas Rogers, Sara Shneiderman, and David Watts laid the foundation for many of the ideas explored in this book. Josh Rubin, the other half of the "terrible twosome," has been a staunch supporter, thoughtful critic, confidant, and dear friend since our first year of graduate school. Michael Degani's steadfast friendship has been a source of intellectual and emotional nourishment. Aniket Aga, Anne Stefanie Aronsson, Samar Al-Balushi, Sarah Le Baron-Von Bayer, Annie Claus, Luisa Cortesi, Madhavi Devasher, Shawn Fraistat, Isaac Gagnè, Shaila Seshia Galvin, Dana Graef, Sahana Ghosh, Annie Harper, Shafqat Hussain, David Kneas, Min Liekovsky, Minhua Ling, Atreyee Majumder, Sarah Osterhoudt, Ellen Rubinstein, Ryan Sayre, Nathaniel Smith, Vikramaditya Thakur, and Jun Zhang were not only scholarly interlocutors but also kept (and continue to keep) me going with their friendship. It seems fitting that Tariq Thachil and Piyali Bhattacharya were there in the beginning and also for the end of the book; our year of living together, with its unceasing love and laughter, has birthed a friendship that has taught me a great deal about what it means to make family. Rohit Naimpally and Naina ensured that my last year in New Haven never lacked for warmth, companionship, and ice cream.

I worked on parts of this book while at the University of Illinois, Urbana-Champaign, where an exceptional group of colleagues and students offered thoughtful insights and support. In particular, I would like to thank Tariq Omar Ali, Ikuko Asaka, Andrew Bauer, Jessica Greenberg, Onni Gust, Marc Hertzman, Rana Asali Hogarth, Charles Roseman, and Behrooz Ghamari Tabrizi for their warm friendship and lively and thoughtful feedback on my work. I am also grateful to Jessica Vantine Birkenholtz, Trevor Birkenholtz, Antoinette Burton, Jodi Byrd, Jane Desmond, Vince Diaz, Virginia Dominguez, Brenda Farnell, Samantha Frost, Susan Koshy, Trish Loughran, Korinta Maldonado, Martin Manalansan, Ellen Moodie, Robert Morrissey, Shankar Nair, Andrew Orta, Jesse Ribot, Gilberto Rosas, Jeongsu Shin, and Roderick Wilson for clarifying and pushing my ideas further. Sadly, Nancy Abelmann passed away before she could see this book come to fruition. She was a constant source of support, friendship, and inspiration, and her many careful readings of my work improved it immensely.

I could not have asked for a more socially congenial and intellectually challenging setting to finish this book than the University of Washington. My colleagues, students, and friends here in Seattle have been a remarkable fount of wisdom and inspiration, and their advice has been crucial in helping me revise the book. In particular, I would like to thank Sareeta Amrute, Megan Carney, Marieke van Eijk, Maria-Elena Garcia, Jenna Grant, Sunila Kale, Celia Lowe, Christian Novetzke, and Priti Ramamurthy for conversations about and readings of the book that have helped to illuminate new directions of thought and clarify old ones. I am also grateful to Kemi Adayemi, Ann Anagnost, Daniel Bessner, Manish Chalana, Rachel Chapman, David Citrin, Jean Dennison, Vanessa Frieje, Julian Gantt, Sara Gonzales, Jenna Grant, Daniel Hoffman, Judy Howard, Tony Lucero, LeiLani Nishime, Dan Paz, Michael Vicente Perez, James Pfeiffer, Chandan Reddy, Josh Reid, Cabeiri Robinson, Rick Simonson, Amanda Snellinger, Keith Snodgrass, Janelle Taylor, Joy Williamson-Lott, Anand Yang, Megan Ybarra, and Hamza Zafar for their support and friendship.

I am grateful also to a number of friends who do not fit neatly within the bounds of this geography but have supported and nourished me at a distance. Many thanks to Mona Bhan, Juno Parreñas, Benjamin Siegel, Noah Tamarkin, Anand Vivek Taneja, Anand Vaidya, and Bharat Venkat for their friendship and counsel over the years.

I have benefited immeasurably from the generosity and astute insights of numerous other people who have read all or parts of this book over the years or responded to talks and presentations based on it. Sareeta Amrute and Anand Vivek Taneja read the whole manuscript and offered thoughtful and generous comments and suggestions that have improved the book tremendously; in addition to their intellectual generosity, their constant support and encouragement as I finished and revised the manuscript was invaluable. I am also grateful to Sarah Besky, Naisargi Dave, Ann Gold, Christophe Jaffrelot, Hayder-al-Mohammad, Anand Pandian, Juno Parreñas, Harriet Ritvo, and Tariq Thachil for their advice on improving particular sections of the book. I would also like to thank Daud Ali, Daniela Berti, Janet Brown, Matei Candea, Stephen Dueppen, Colter Ellis, Shelley Feldman, Thomas Blom Hansen, Michael Hathaway, Annu Jalais, Jake Kosek, Ashish Koul, Mekhala Krishnamurthy, Philip Lutgendorf, Karen Mager, Nayanika Mathur, Tapsi Mathur, Lisa Mitchell, Geeta Patel, Nancy Lee Peluso, Sarah Pinto, Mubbashir Rizvi, Tanika Sarkar, Uditi Sen, Samira Sheikh, Alistair Sponsel, Ramya Sreenivasan, Tony Stewart, Ajantha Subramaniam, Yuka Suzuki, Sharika Thirangamma, Emiko-Ohnuki Tierney, Mudit Trivedi, Anand Vaidya, Bharat Venkat, Sharon Wilcox, Michael Wise, and Rebecca Jane Woods for having provided, at various stages of my career, insightful comments on the ideas that are discussed in the pages that follow. I shared parts of the book in talks at various conferences, workshops, and universities over the last few years, and I am grateful to the audiences in these multiple settings for their comments.

Support from the Program in Agrarian Studies, the Yale Center for South Asia, the MacMillan Center at Yale, and the Social Science Research Council (SSRC) Dissertation Proposal Development Fellowship Program made the early research for this project possible. The American Institute of Indian Studies Junior Fellowship and the International Dissertation Research Fellowship from the Macmillan Center funded the main fieldwork for the project between 2010 and 2011. I would like to offer my gratitude to the AIIS for honoring this book with the Edward Cameron Dimock, Jr. Prize in the Indian Humanities. I thank the University of Washington Helen Riaboff Whiteley Center for providing, on two separate occasions, such a serene and beautiful environment in which to write and revise the book. I would also like to thank WIRED for making my first stay at the Whiteley Center possible.

Portions of the book have been published before: portions of chapter 2 were published as "The Goat that Died for Family: Animal Sacrifice and Interspecies Kinship in India's Central Himalayas" in *American Ethnologist*, and portions of chapter 4 were published as "Monkey Business: Macaque Translocation and the Politics of Belonging in India's Central Himalayas" in the *Comparative Study of South Asia, Africa, and the Middle East.*

I thank Jane Desmond, Kimberley Marra, and Barbara King, the editors of the series Animal Lives, for their faith in and encouragement of this project. I am also grateful to my editor Douglas Mitchell at University of Chicago Press for his expert advice and to Kyle Adam Wagner, also at the press, for his tireless support. Thanks to Steve LaRue for his precise and thoughtful copyediting. Thanks also to the rest of the team that worked hard to produce this book: Skye Agnew, Adeetje Bouma, Mary Corrado, Ashley Pierce, Julie Shawvan, and Levi Stahl. I could not be more appreciative of the two excellent readers of this manuscript. One was anonymous and the other, Naisargi Dave, identified herself to me. Their reviews were generous, insightful, and precise. Their comments helped me to clarify and push the arguments in the book further, and it is immeasurably better for their careful attention. Thanks also to Philip Lutgendorf for his thoughtful comments on the book proposal and two chapters of the book at an earlier stage.

There are many people to thank in India. Anisha Chugh has been an indefatigable supporter of numerous projects over the years and has brought much love, laughter, and joy to my life. I am grateful to Ritika Goswami, Manusmriti, Tapsi Mathur, and Yamini Vijayan for their steadfast friendship and support without which the last two decades would have been much grayer. Thanks also to Mahesh Rangarajan for his encouragement, insight, and friendship over twelve years; he has shaped this project in critical ways. My teachers at the Center for Historical Studies at Jawaharlal Nehru University, especially Neeladri Bhattacharya and Tanika Sarkar, have been critical to shaping the ways in which I think. I would not be here without their guidance. I am grateful to Rajesh Thadani, V. K. Madhavan, and the staff of Chirag (Central Himalayan Rural Action Group) for their tireless support of this research and for making it possible in the first place. I am also indebted to Rajesh Singh, Deepak Rawat, Jai Raj and other members of the Forest Department for the assistance they provided me during my research. Thanks also to the staff at the National Archives of India in Delhi and the Uttar Pradesh State Archives in Lucknow.

I thank my parents-in-law, Rani and Krishna Athreya, for their love and good wishes. My sister Charu believed in me every step of the way, and her support has been critical in sustaining my work. My parents, Kannan and Nalini Govindrajan, have shared every happiness, doubt, and sorrow that emerged over the course of the research for and writing of this book. For the unconditional love and security they have always offered me, I dedicate this book to them. Sheru, Devrani, Lobo, Tuffy, Chunu, Tipton, and Mika have brought me immense delight with their creature companionship and have compelled me to think about multispecies relatedness in new ways.

Finally, Jayadev Athreya has eaten, slept, and breathed this project for as long as I have, and his insights and questions shape every single page. He has helped me understand the joy that can ensue from being open and vulnerable to another. I look forward to a lifetime of more learning and adventure with him.

Introduction

In September 2010, only a few months after I had started immersive fieldwork in the Central Himalayan state of Uttarakhand in India, Kumaon—the eastern part of the state—was ravaged by several days of furious, unremitting monsoon rain. Boulders tumbled onto the highway; trees fell to the ground with thunderous cracks that made those of us cowering in our houses fear that the sky had burst open; the weight of collapsing mudbanks pushed several houses into the ground. As the rain continued to rage, people readied themselves for the onerous repair and reconstruction that would follow. When the clouds were finally spent, Chirag (Central Himalayan Rural Action Group), a local NGO, asked me to accompany their fieldworkers on a tour of nearby villages where they would assess the damage in order to assist the state in its relief efforts. Nimmi *di*, the Chirag worker whom I was accompanying, decided that we would start in Simal village, where several houses had been damaged by the rain.[1] The rain had transformed the undulating landscape into a breathtaking patchwork of different shades of green. Nimmi *di* admired the view as I stopped to take pictures and then instructed me to tuck the ends of my *salwar* into my socks as she was doing. Our path through the village, she warned, was infested by leeches, and she had forgotten to bring the salt that she had wrapped up in a torn sheet of newspaper earlier in the morning.

Our first stop was at Munni Rekhwal's house, where a retaining wall had collapsed, burying the kitchen and back rooms in an avalanche of sludge. Munni, a widow, lived alone. When her house was destroyed, she had moved briefly into a neighbor's house before inexplicably returning to take up quarters in the only room of her house that had not collapsed. Nimmi *di*'s task was to convince her to move to the safety of a relative's home until the house could be repaired. When we got there, we found Munni in the *goth* (shed), standing beside Radha, a tawny Jersey cow with dark and expressive almond-shaped eyes ringed with thick lashes. When my eyes adjusted to the darkness, I saw that she was little older than a calf, no more than two or three years old. All the other animals—several goats and another cow—were tethered to stakes driven into the ground outside. The air in the *goth* was scented with pine and *gobar* (cow dung), a heady and strangely pleasant fragrance that would linger on my clothes and in my hair all day. Standing beside Munni, a good foot taller than her, was a *jhaadnewallah*, a healer, who repeatedly stroked the cow's trembling back with the wispy branch of an acacia while murmuring mantras under his breath. His lips moved rapidly, but I could hear nothing but the heavy sighs of the cow. Radha, we learned, had refused food and water for the last two days. The *jhaadnewallah* left ten minutes after we got there, telling Munni to rub the juice of a lemon inside Radha's mouth. He would be back tomorrow if her health did not improve. As Radha sank into the squelching, soggy ground with a loud groan, Munni began to carefully arrange large clumps of dried pine needles around her for warmth. Nimmi *di* and I walked into the *goth*, trying to avoid the glistening clumps of dung that clung to the bed of pine needles and oak leaves that Munni had laid for her animals. Tucking her *kurta* under her knees before squatting on the ground next to Munni, Nimmi *di* tried every trick in her arsenal to persuade the widow to move to a safer location. I added my entreaty to hers, but Munni's attention was focused entirely on Radha, whose nose she caressed lovingly while alternately beseeching and ordering her to eat some grass. Though she did not eat, Radha responded to Munni's overtures by nestling her head in the crook between Munni's neck and shoulder.

I was not surprised when Munni flatly refused to move. Nothing in her demeanor had suggested that she was moved by our pleas. Nimmi *di* was nevertheless visibly frustrated. "Why will you not move?" she asked Munni. "What if a leopard attacks you at night? You don't have a back wall."

"Let the leopard come," Munni said, with frustrating calm. "I'll see what to do then. Look what happened when I left for one night. Radha took ill."

"The rain might have caused her illness," Nimmi *di* suggested exasperatedly. "The roof of your *goth* was damaged by the rain, wasn't it? Maybe she has a fever. Why not call the doctor?"

But Munni was firm:

It's not that. She [Radha] knows something bad has happened. Every evening I give the animals some grass. But that night I stayed elsewhere so Jeevan [her neighbor] gave them grass instead. I was so sad. It was the first time that I had not fed the animals myself. I thought "why did this happen to me?" I cried.

The next morning, Jeevan told me Radha had cried all night. When I went to the *goth*, I saw that she hadn't touched the grass. Usually she eats her mother's share as well. It's been two days; she hasn't eaten anything. Now if I try to go anywhere she makes such a racket that I have to come back. Mahender (the *jhaadnewallah*) says that someone may have cast the evil eye on her. But I don't think so; she doesn't even give milk yet. She is sad (*dukhi*) because of what has happened. She knows when I am anxious (*pareshan*), and she also becomes anxious. Such is *moh-maya* (love). How can I let her die while I stay elsewhere for my own comfort? I will stay here. So what if my family is not here? These animals are my family too.[2]

On Relatedness

What do we make of Munni's refusal to prioritize her safety over Radha's health? As we commenced the steep climb to the next household we were visiting, Nimmi *di* vented her irritation at Munni in between great gasps of air. "These village women are so stubborn," she muttered. "Imagine staying in that crumbling ruin of a house because of your cow. May I never lose my mind like that. Did you see her rubbing dried *gobar* off Radha's back with her *saree*? Rather a mad [*pagal*] love, don't you think?" I ventured that if, as Munni had argued, her well-being was entangled with that of her animals, then perhaps her decision to remain with them made sense. Since their lives were so intimately connected, Radha's unhappiness and subsequent ill health had emerged from the misfortunes that had befallen Munni; the destruction wrought by the rains had touched the lives of everyone in this multispecies family, and the only way for them to recover was together. Nimmi *di* threw a skeptical look my way as I shared this thought with her. "Has one Radha fallen in love with another?" she asked me teasingly.[3] A few minutes later, she conceded that perhaps she did not understand this *rishta* (relation) because her life was not tied to the lives of animals in the same way as Munni's. When such knots (*bandhan*) are tied, she said, it is difficult to disentangle them. Whether good or bad, life can now be lived only through these ties that bind one to another.

What does it mean to live a life that is knotted with other lives for better or worse? How do such knots come to be tied? In this book, I am interested in exploring how these knots of connection produce a sense of relatedness between human and nonhuman animals. I use the concept of relatedness (Carsten 2000) to capture the myriad ways in which the potential and outcome of a life always and already unfolds *in relation to* that of another.[4] To take these entanglements as constituting forms of relatedness is to acknowledge that one is not formed as a self in isolation but through the "doing and performing" of relations—both desirable and undesirable—with a host of other beings whose paths crisscross one's own in ways that defy the integrity of bodies and communities (Schneider 1984, 165–66). To think of these everyday relations as creating a terrain of relatedness is to question, as generations of feminist scholars have done, the "naturalness" of categories such as kinship and biology, nature and culture, sex and gender, human and animal, and to think instead about the particular contexts of their naturalization. It is to recognize that human pasts, presents, and futures are gathered with the pasts, presents, and futures of the multiplicity of nonhuman animals who share worlds with them. When Munni said that her anxiety was felt acutely by Radha, a knot of connection between them that emerged through the daily intimacy of care, it was precisely this entanglement of fates and consciousness to which she was referring.

The relatedness I explore in this book is decidedly uninnocent. To be related to another is to be imbricated in their making even when one is indifferent to, disgusted by, or hostile to them. Mutuality and connection do not imply an erasure of difference or hierarchy. As Donna Haraway (1991, 177) reminds us, if "two is too many" to describe the being who is made through relations with others, then "one is also too few." Relatedness emerges from the connections between individual animals that are akin to the "partial connections" that Marilyn Strathern (2004, 39) describes among people in Papua New Guinea in that they create "no single entity between them" and, indeed, rely on some degree of (hierarchical) separation between different figures. The existence of difference is critical to relatedness because, as Eduardo Viveiros de Castro (2012, 38) notes, "kin can only be made out of others." Over the rest of this work, then, I hope to show that multispecies relatedness draws as much on incommensurable difference as ineffable affinity between particular individuals for its emergence.

In what follows, you will read about relatedness between humans and animals as it emerges through the ritual sacrifice of goats, an act of violence that is excoriated by animal-rights activists, and has increasingly become the subject of legislation; right-wing political and religious projects of cow protection that are troubled by the fact that the bodies of cattle are too wayward and distinct to be contained within a stable and homogenous symbol; a contemporary politics of exclusion and belonging that has been sparked by the sudden and unwelcome appearance of monkeys translocated from cities to mountain villages where they feed circulating anxieties about the erosion of cultural identity; the conservation of wild boar whose protection by the state is contested by villagers on the grounds that the history of these animals' wildness is fluid and contingent; and stories about bears who are believed to abduct and have sex with women, a tale of queer crossing that blurs the boundaries between species. Each of these chapters will trace a different form of relatedness, paying particular attention to how it is shaped by encounters between different animals across species and what kinds of material and affective labor that engagement entails. In the epilogue, I will turn to what the violent relatedness between leopards and the dogs they eat can illuminate about the intersections of violence, love, and everything in between that constitutes the stuff of relatedness.

There are several reasons why relatedness is a compelling concept with which to understand the complicated entanglements of people and animals in Kumaon. First, the idiom of relatedness was often invoked by people in describing how their lives were entangled with animals. Recall that Munni used the word *parivar*, or "family," to describe the relationship she shared with her cows and goats. As we shall see in chapters 2 and 3, many others also used kinship terms drawn from genealogical models of kinship to describe similar relationships of intimacy, attachment, and love that exceeded the biological and genealogical. The language of kinship was invoked by people even when relatedness with animals was unwanted and produced harmful effects in the lives of everyone involved. My friend Jagdish, who farmed a quarter of an acre, complained that his daily battles with rhesus macaques translocated from north Indian cities to rural areas in the mountains appeared as scenes out of the *Mahābhārata*. "Here too, the battle starts at dawn and ends at dusk. Our mountain monkeys are like Hanuman. These outsider monkeys, they are Kauravas."[5] The evocation of kin categories from the epic *Mahābhārata* to describe everyday relationships with macaques was significant. Whether ally or enemy, the monkey was a relative. Desirable relatives were welcome to stay, but undesirable relatives—the monkey interlopers from cities—would have to be met in daily battle.

In taking these situated conceptions and enactments of relatedness seriously, I draw inspiration from the work of Donna Haraway (1991, 2003, 2008, 2016), whose writing on cyborgs and companion species reminds us that it is not just humans who are composed relationally but all critters. "Kin making is making persons," she writes in her latest book *Staying with the Trouble*, but "not necessarily as individuals or human. Kin is an assembling sort of word" (2016, 103). Haraway also reminds us that the critical challenge confronting feminist scholarship in this moment is that it must "unravel the ties of *both* genealogy and kin, and kin and species" (2016, 102, emphasis mine). While the link between genealogy and kin has been productively troubled over the last few decades, feminist scholarship on kinship and relatedness, as Haraway points out, has, until recently, remained largely focused on exploring human realms.[6] In this book, I engage this missing piece by attending to how relatedness always and already exceeds the human in its everyday doings in Kumaon. My approach to relatedness is firmly grounded in Kumaoni villagers' conviction that kin making is a multispecies affair. When people in Kumaon said that they were related to animals, the metaphors of kinship were grounded in what they saw as truth. For them, to be formed as a person was to be formed *relationally* with other animal persons. In letting people's theories about their own lives guide my ethnographic practice, I am guided by the belief that a decolonized anthropology is one that must not only seriously engage but also be driven by the multiple points of view of its interlocutors (Nadasdy 2007; TallBear 2011; Biehl 2013).

My second reason for framing interspecies mutuality as relatedness is that the experience of sharing connection with and attachment to other animals was not restricted to humans alone. After years on end of living with the actual animals who inhabited this landscape, some of whom were not only ethnographic subjects but friends and family, I could not deny that they responded to and offered up their own gestures of relatedness in the course of quotidian relationships with one another and humans. Throughout this book, I am committed to rendering animals as I encountered them—not as a symbolic foil for human representation but as subjects whose agency, intention, and capacity for emotion was crucial in shaping the relationships they made with humans. Whether Radha, who responded to Munni's absence by taking ill, the goat in chapter 2, who treated the woman who had helped birth her as a second mother, the monkey in chapter 4, who left her troop to seek out the companionship of an old man, or the pig in chapter 5, who is in a "domestic but not domesticated relationship" with his owner, the animals whose lives were imbricated with those of humans would often convey their sense of relatedness to particular individuals, both human and animal.[7] I am not suggesting that all animals felt a sense of intimate connection with humans or even that different species or individuals within a species experienced relatedness in the same way. Instead, throughout this book I chart how relatedness—the relational unfolding of life—was expressed and experienced in varied ways by different animals along the continuum in the course of their fleshy entanglements with one another. Instead of focusing on human entanglements with a single species, this book follows the lives of a variety of animals across different species, allowing it to complicate and disaggregate the all-too-capacious category of animal in ways that permit a recognition of the multiplicity and diversity of experiences and subjectivities among different animals. The animals I engage in the pages to follow include people, goats, cows, monkeys, wild boar, bears, leopards, and dogs, a veritable bestiary that is faithful to the abundance of entangled relationships I encountered during my fieldwork.

Inhabiting the Terrains of Relatedness

The forest-edge villages where I conducted research are located in Kumaon, the eastern *mandal* (administrative division) of Uttarakhand that is bordered by Tibet, Nepal, and the Indian state of Uttar Pradesh; Garhwal, a historically, linguistically, and culturally distinct area, forms the western *mandal*. Located between 6,500 and 7,800 feet above sea level, the villages I lived in straddled a series of ridges running east and west that separate the Himalayas from the Bhabhar and the Terai, the tracts that lie just south of the Shiwalik hills. Cultivation—difficult and unrewarding work in this terrain—was practiced on a rather modest scale in these villages and, increasingly, was abandoned altogether by younger villagers who were no longer content to seek their future in the soil. However, most villagers continued to keep livestock animals— overwhelmingly cows and goats with the occasional buffalos or chickens where I lived—for milk and meat, as well as cats and dogs to guard the house from intruders big and small, whether mice, snakes, or other humans.[8]

These animals lived in intimate proximity to their human caregivers. In a few of the older stone and wood houses, livestock animals still lived in rooms on the ground story with people's rooms perched directly above them. An older woman whose back was perpetually bent after years of hard labor told me that she had stubbornly resisted her children's efforts to move the *goth* a little distance from the house because she would miss the bodily heat and sweet fragrance of her cows wafting up through cracks in the wooden floor-boards even if it meant having to tolerate the mosquitos and other insects that accompanied it. In homes where livestock animals had been moved to their own shed, this relationship based on mutual residence lingered in semantics; rooms on the lower level of the house were still referred to as *goths*, remind-ing everyone of their previous occupants. To live in the village, then, entailed learning how to live with these different animals. It involved the recognition that the rhythms of one's day and, indeed, one's life were caught up with those of another's. In chapters 2 and 3 and then again in the conclusion, I describe how these daily proximities and entanglements produced a deeply felt sense of mutuality between villagers and livestock animals that was predicated on their imbrication in one another's lives.

A dense mixed forest of oak, pine, rhododendron, and other species, more than a few thousand acres of which was controlled by the Indian Veterinary

FIGURE 1. Most villagers kept cows and goats and lived in intimate proximity to them.

Research Institute (IVRI)—an institution set up in the late nineteenth century by the colonial state with a view to developing prophylactic measures for infectious animal disease—abutted these villages.[9] In this reserved forest, where hunting was illegal and the collection of fuel and fodder by villagers severely circumscribed by the state, the paths of people and the animals who lived in the village alongside them crisscrossed the paths of other animals for whom the forest was home—deer, monkeys, wild boar, leopards and other felines, to name but a few of the creatures who roamed these woods.

Some of these animals, as this book describes, moved between the "forest" and the "village," shifting physical and conceptual spaces themselves, with an ease that made it difficult to sustain human "placements" of them as creatures of the forest distinct from creatures of the home (Philo and Wilbert 2000; Govindrajan 2015b). In what follows, I trace how animals' refusals to stay "in place"—whether monkeys translocated from cities who declined a life in the forest or leopards who created a niche in and around villages for themselves—produced fertile ground for encounters that forced people to recognize that their different lives had come to be inextricably connected in ways that made it difficult to sustain the fiction that humans belonged in one place and wild animals in another no matter how deeply people wanted that fiction to be true.

What, other than the proximity that life in these spaces produced, fostered the entanglements that were recognized as constituting different modes of relatedness—family, friendship, enmity, and indifference? Relatedness could encompass bonds of kin rooted in mutual nurturance that took time and work, like the relationship between Munni and Radha. The kin-like relationship between them emerged through daily, arduous—and gendered—practices of care and labor that led them to be open to the joy, anxiety, and suffering of the other. Their relatedness thus emerged out of the "intersubjective phenomenology of care" (Livingston 2012, 112). Their connection, which Munni characterized as distinct from genealogical kinship—*so what if my [real] family is not here?*—but no less affectively powerful—*these animals are my family too*—speaks to the ways in which multispecies relatedness, like human kinship, exceeds the biological even if humans continue to rely on genealogical idioms of kinship to describe it.

The horizon of relatedness, however, was not circumscribed by kinship born from nurture. In Kumaon, as in other parts of South Asia, anthropologists have chronicled how relatedness is recognized as an effect of "shared mutual substance" imbibed through constant transactions between beings and the landscape they inhabit (Sax 2009).[10] To eat from the same soil, to drink from the same rivers, and to worship the same gods, in other words, is to be related; the "substance of kinship" is acquired over time through such exchanges and entanglements (Carsten 1995). In the mountains, many described this relatedness between humans and other animals as further enriched by the common experience of living in and being subject to the rules of the *devbhumi*—the land of gods, as Uttarakhand is often called in deference not only to the high gods of Hinduism whose shrines are located in the upper reaches but also to the thousands of local deities (*devi-dyavtas*) who reside in its mountains and are associated with its stones, streams, and forests. *Pahari* (literally, "of the mountain") animals, I was told on multiple occasions, were related to *pahari* people by virtue of their shared subjection and relatedness to *pahari devi-dyavta*. Thus, a *pahari* goat would understand the need for his own ritual sacrifice to a local deity in a way that a goat from the plains simply could not (Govindrajan 2015a).

Similarly, a *pahari* leopard's predation on people could, in some cases, be explained by the fact that he or she was acting on a deity's instructions to punish recalcitrant human devotees (Govindrajan 2015b). In her rich exploration of relatedness between humans and tigers in the Sundarbans, the anthropologist Annu Jalais (2010, 74) argues that "the reason why the Sundarbans forest fishers believe they are tied in a web of relatedness with tigers is because they have the same symbolic mother in Bonbibi, because they divide the forest products between themselves and tigers, and because ultimately they share the same harsh environment, which turns them all into irritable beings." In

FIGURE 2. The landscape of the region is dotted with thousands of humble shrines dedicated to local *devi-dyavtas*.

Kumaon, too, the actions of *pahari* animals could often be rendered comprehensible to humans because they occurred within a framework of mutual relatedness that encompassed not just humans and animals but also the gods. I am not suggesting that "shared substance" eroded any and all sense of individualism but simply that many people believed that some moral and spiritual qualities could be shared even across difference.[11]

Knots of connection, however, were not always nurturing or heartwarming. As Carsten (2013, 246) points out, ideas about mutuality and care tend to dominate studies of relatedness that, on the whole, give off a "warm, fuzzy glow rather than a cold shiver." This focus on the positive aspects of relatedness, she argues, obfuscates how kinship is shaped by and enables "differentiation, hierarchy, exclusion, and abuse." Indeed, the challenge that confronts critical feminist kinship studies, then, is not just to denaturalize relatedness by extending its experience and practice to nonhumans but also to expose and undermine the circuits of race, gender, and species difference and violence within which it takes shape. With that in mind, this book is guided by Lauren Berlant's (2011b, 687) call for theories of attachment that engage not just the "ambitions and capacities of love . . . [and] other modes of relating . . . the ones involving proximity, solidarity, collegiality, friendship" but also the "hatreds, aversions, and not caring."

Multispecies relatedness in Kumaon was often a rather knotty affair. The fellowship between *pahari* humans and animals was rooted not just in shared substance or mutual devotion to *devi-dyavta* but also in the common experience of inequality and exploitation. In chapter 4, I describe how villagers claimed a kinship with *pahari* monkeys on the grounds that they had both experienced displacement from their homes by outsiders. To be related as *paharis*, then, was to be related through a shared history of neglect and exclusion. Further, appeals to claim certain beings as kin and denounce others as enemies were often rooted in violent, xenophobic politics that drew sharp boundaries around those who were to be included within community and nation. In chapter 3, I explore attempts by right-wing Hindu nationalists— supported by their political arm, the Bharatiya Janata Party (BJP)—to constitute a pure and unified community of Hindus united around the symbol of the mother cow. This is a project that has its roots in the nineteenth century, and relies on the portrayal of Muslims, Christians, and Dalits as a threat to the cow and, by extension, to an imagined community of her upper-caste Hindu children. I describe how many villagers viewed such attempts to impose kinship between them and cows at large as an act of coercion. However, such coercive framings of kinship did not go unchallenged; as I show, villagers responded to them by imagining alternate forms of relatedness that emphasized *pahariness* and *pahari* Hinduness over Hinduness at large as the grounds of connection between themselves and their cows. I argue that these alternate forms of relatedness drew on the fact that the distinctive characteristics and tendencies of individual cows were incommensurate with the homogeneous metaphor of the mother cow. Relatedness, in other words, was crucially shaped by the liveliness and unpredictability of the animals in question.

The relatedness I describe in the rest of the book was also a product of the complicated entanglement of race, colonialism, caste, and animality. As Frantz Fanon (1963, 42) observes, the Manichean nature of colonial discourse reaches its logical conclusion in the "dehumanization of the colonial subject." "In plain talk," Fanon declares, the colonized individual "is reduced to the state of an animal." Indeed, in colonial India, the "very notion of the native belonged to the grammar of animality" (Mbembe 2001, 236); the humanity of the colonizer was contingent upon the racialized animality of the colonized. From the late nineteenth century onward, colonists argued that this irrepressible tendency toward animality in the colonized subject made it necessary for the colonial state to protect nonhuman "game" from the beast-like savagery of the native. This was accomplished by reserving the right to hunt for a select handful of colonists. In chapter 5, I trace how the "wildness" of the native was and remains crucial to projects of colonial and postcolonial conservation. Yet I argue that the category of wildness is not entirely lost to its colonial inheritance. In order to disentangle the complicated threads of wildness, chapter 5 follows the story of an experimental pig who escaped confinement at the IVRI in the 1960s, and made her refuge in the surrounding forest. This runaway sow is believed by villagers to be an ancestor of the wild boar who swarm these forests today. I argue that her history, even as it bears indelible traces of racial meaning and the workings of sovereign colonial power, contains within it the potential for what I call an *otherwild*, a messy wildness that reconfigures, unsettles, and exceeds the ways in which the "wild" and the "animal" buttress intersecting projects of colonial, caste, and species difference and power.

In the book I explore how such relations of similarity and difference are enacted on an increasingly contested ethical and political terrain. I was drawn to the villages where I conducted fieldwork because they offered a glimpse into a multiplicity of attachments between very different kinds of animals in a setting that was not as firmly bounded or specialized as the factory farms, slaughterhouses, or conservation refuges on which multispecies ethnographers usually tend to focus. While studies of such distinctive contexts and interactions are crucial, multispecies relationships in relatively ordinary sites such as parts of rural Kumaon have been largely overlooked. Studying everyday practices of relatedness in rural areas allows us to approach the fundamental questions of multispecies ethnography from a different direction.

It is worth noting that these villages were not primitive idylls in which to track the supposed affinity of simple mountain villagers for nature.[12] Instead, they were dynamic sites of multiple and shifting encounters in which multispecies relatedness and the forms it should take were negotiated by a range of actors—villagers, NGO workers, animal-rights activists, Hindu reformers and right-wing nationalists, and the nonhuman animals who were the subjects of these debates themselves. The "village" in this book, then, does not represent a self-contained, pristine Arcadia outside time and space—the mythical fetish of colonial ethnography—but a porous and malleable world shaped by a long and layered history of engagement with myriad economic, political, cultural, and ideological contexts and conversations.[13]

I emphasize this because proponents of what has been called the "ontological turn" sometimes suggest that native conceptions of the nonhuman world constitute a radical alterity, a difference they sketch in totalizing and homogenous ways that bear troubling resemblance to the idea of culture as a bounded whole.[14] However, as Judith Butler reminds us, the question of who and what is considered real is a question of power, and "power often dissimulates as ontology" (Butler 2004, 215). With that in mind, this book departs from this body of work in its rejection of the notion of a stable alterity among particular people and in particular places (cf. Bessire 2014; Allen and Jobson 2016). Instead, I pay close attention to how the stuff of relatedness grows only thornier as it is drawn into bitter battles over identity, locality, and morality that have become more urgent over the last decade and a half.

As David Graeber (2015) points out, most people whom anthropologists study would disagree with the proposition that they live in a fundamentally different reality or nature. Few pahari villagers see themselves as proponents of a unique metaphysic or as inhabiting a radically different world than other humans. On the contrary, I argue, their main preoccupation is how to live well and as "modern" subjects in partial, overlapping, and sometimes incoherent worlds that they must share with other humans, nonhumans, and institutions. Theirs is not a radically incommensurate world but one that is "composed" at the juncture of multiple worlds that are constituted by the daily practices of a heterogeneous range of actors and that is subject to the constraints of time and space (Latour 2010). With that said, let us now to turn to a closer examination of how these forms of relatedness shift in relation to contemporary political and ethical projects.

Knotty, Gnarly Histories

The state of Uttarakhand was created as a separate state in the Indian federation in November 2000. The decision to carve out a new state from the "parent" state of Uttar Pradesh was influenced by some combination of political expediency and the pressures exerted by a mass regional movement in favor of political autonomy that gathered momentum in the 1990s after the Uttar Pradesh government proposed legislation that would reserve a chunk of government employment and education opportunities for Other Backward Classes (OBCs); this proposed legislation caused consternation in the hills where the majority of the population was upper caste (Mawdsley 1998). The demand for a separate state was based on the idea that the mountain regions of Uttar Pradesh had been exploited for natural resources by its political leadership, which was overwhelmingly made up of plains people, but had been ignored when it came to the allocation of jobs and development projects. As in the case of Darjeeling, resources that flowed down the mountain did not flow back up (Besky 2013). Protestors lamented the fact that the *pahar* (mountains) had been treated by plains-based administrators and politicians as an "internal colony" (Pathak 1997; Mawdsley 2002); they demanded local control over *jal* (water), *jangal* (forest), and *zameen* (land). The idea that development needed to account for geographical and cultural specificity was at the heart of the call for the creation of a mountain state that would cater, first and foremost, to the needs of its *pahari* populace (Aryal 1994; Mathur 2016; Mawdsley 1998; Rangan 2000). Only a state controlled by *paharis*, it was argued, would be able to bring development to the mountains.

During my several years of fieldwork in Kumaon, I sensed a palpable and pervasive disappointment with the new state and its lack of effects. Development—as a "matter of moral as well as material aspiration"—remained tantalizingly out of reach for most, which only enhanced its appeal (Pandian 2009, 7). Rates of rural unemployment remained high in the mountain districts of the state. As a result, migration from the mountains to districts located in the plains regions of Uttarakhand or to other states continued unabated. The Census of India 2011 revealed that rates of migration from the mountain districts of Uttarakhand had gone up since the last Census of 2001, betraying the hope that the creation of a putative mountain state would better the lives of *paharis* by bringing development to the mountains. There was a widespread sense that the mountains and their residents were in the throes of a crisis—a subject I discuss at greater length in chapter 4. "Paharon me sthithi bahut kharab hai" (the situation in the mountains is terrible) was a ubiquitous refrain in conversations I had with *paharis*, both those who remained in the mountains and those who had left to seek employment elsewhere. The only difference from the 1980s and 1990s, people told me, was that *paharis* were now second-class citizens in their own state (Govindrajan 2015c), their needs and experience subordinated to those of outsiders from the plains who were taking over the mountains through their investments in land, mining, and industry.[15] People constantly expressed the fear that this crisis would erode *pahari* cultural identity because it was not just the mountains that were left behind by migrants but also distinctive *pahari* cultural practices. As lack of opportunity came to define life in the *pahar*, people were increasingly anxious that *pahariness* itself would become synonymous with backwardness, something to be shunned in the embrace of a "modernity" that was always elsewhere.

Such fears have a deep history. The idea that *paharis* are backward, superstitious, and unorthodox in their habits goes back at the very least to the colonial period (Govindrajan 2014; Pande 2015). G. W. Traill, the first commissioner to govern the region after the British vanquished the Gorkhas and took control of Kumaon in 1815, noted that "an attempt to collect the numerous superstitious beliefs current in these hills . . . would be an endless task, the result of which would by no means repay the labor bestowed, as these beliefs are for the most part rude and gross, displaying neither imagination nor refinement in their texture."[16] The anthropologist Gerald Berreman, who conducted fieldwork in Garhwal in the 1950s, noted that these beliefs about the backwardness of *paharis* were shared by people from the plains. "Paharis are considered to be rustic, degraded Hindus by most plainsmen," he wrote in his classic *Hindus of the Himalayas*. "Plains Brahmins and Rajputs often reject the caste status claims of their Pahari caste-fellows, largely because of their unorthodoxy" (Berreman 1963, 139). The unorthodox practices for which they were shunned by plains people, Berreman notes, included animal sacrifice, the frequent consultation of diviners and shamans, and a general failure to observe the ceremonies and restrictions that were held up as essential by high-caste plains people. *Paharis* thus occupied the "savage slot" in this colonial hierarchy of virtue and progress (Trouillot 1991). The colonial legacy of identifying the mountains as a site of backwardness and apathy to which development must be brought from the outside continued to resonate in the region during my time there. Some *paharis*, especially young men and women who were keen to cultivate "modern" selves, echoed such discourses, locating their lack of development in their inability to leave practices such as animal sacrifice in the "primitive" past where they belonged.

Very different social organizations jumped into the fray to reiterate this connection between *pahari* cultural practices and the lack of development of the *pahar*. One such organization, whose intervention into these debates I track throughout this book, is People for Animals (PFA), India's largest animal-rights organization, founded by Maneka Gandhi, a well-known politician from one of South Asia's most recognizable political dynasties. Currently a minister of the Indian Parliament from the BJP, which came to power in a general election in 2014, Gandhi is arguably India's most prominent animal-rights activist.[17] In Uttarakhand, the PFA was helmed by Gauri Maulekhi, a trusted associate of Gandhi's. During the course of my fieldwork, I encountered the PFA in a diverse array of arenas—campaigning actively for a ban on animal sacrifice, a process I describe in chapter 2; aiding Hindu nationalists in the enforcement of a ban on cow slaughter, a subject I touch on in chapter 3; and agitating against any efforts to cull monkeys or wild boar, themes that emerge in chapters 4 and 5. In chapter 2, I argue that while the organization was committed to ensuring the flourishing of animal life, it framed its acts of animal rescue in terms that were familiar from other projects of rescue in colonial and postcolonial India—as acts of liberating speechless victims from barbaric oppressors rooted in a savage, superstitious tradition.

Perhaps it is the weight of such tangled histories that can explain the alliances the PFA built in the course of its activism. Consider, for example, their campaign against animal sacrifice in which they were joined by the Arya Samaj and All-World Gayatri Pariwar, Hindu reform organizations that are committed to purging Hinduism of superstitious deviations and returning it to a pure and original Vedic religion. This is a tradition of religious reform and revivalism that goes back to the nineteenth century.[18] Animal sacrifice, as I note in chapter 2, has been a particular thorn in the flesh of these organizations who believe that the practice made its way into Hindu practice from Islamic and Christian theology. When reading the Old Testament, Dayanand Saraswati, the founder of the Arya Samaj, was agitated by its advocacy of animal sacrifice. He wrote, "How wild and savage-like it is for God to receive the sacrifice of oxen, and sprinkle blood on the altar. When the God of the Christians accepts the sacrifice of oxen, why should not his devotees fill their bellies with beef and do harm to the world?" (Jones 1992, 58–61). Dayanand's excoriation of sacrifice as a damaging superstition characteristic of Abrahamic faiths lingers into the present. In an article penned for the *Kathmandu Post* during the height of the PFA's 2010 campaign against animal sacrifice in Uttarakhand, Maneka Gandhi claimed that she was often asked why Muslims were still allowed to sacrifice animals. What she responded with was this: "you can't pervert Hinduism into a pale shadow of Islam. . . . According to their religion they are supposed to sacrifice to God, that which they love the most. They have perverted that into buying and killing goats." The emphasis on distinguishing "true religion" from "false and backward superstition" in these two tracts against sacrifice, written almost 135 years apart, illuminates the enduring legacy of colonial discourses of moral improvement that have powerful effects in the present. When seen through this civilizing eye, the performance of sacrifice is the performance of backwardness, a sign that the material poverty of the hills mirrors its "cultural poverty" (Fiol 2013).

In the book, I trace how the weight of these older and broader projects of reform, carried into the present through these contemporary programs of moral development that seemed initially to engage very different sets of concerns, compelled many villagers to engage with the difficult question of what it means to live an ethical life in relation to animals. In the chapter on animal sacrifice that follows, I describe how people imagined and enacted an ethics that emerged both through and despite these shifting imaginaries of violence, suffering, and responsibility. Through engagement with such programs of reform, they emerged as different kinds of ethical subjects than desired by reformers. Rather than aspiring to a universal moral horizon, they imagined and lived an ethical relatedness that took into account the contingencies of "situated" and historical entanglements with animals (Haraway 1988). For them, "questions of ethics and justice," to quote Karen Barad (2012, 69), were "always already threaded through the very fabric of the world." Throughout the book I explore how, in drawing inspiration for ethical conduct in the everyday knots of relatedness that drew them into a web of mutuality with animals, villagers

FIGURE 3. "If you must sacrifice, then sacrifice your inner evils (alcohol, gambling, theft, smoking, and tobacco)." An antisacrifice billboard issued by the Arya Samaj to coincide with the Naina Devi festival in Nainital in 2010.

were able to shake off the yoke of a (post)colonial imaginary that located backwardness and savagery in their very nature and therefore in their pasts, presents, and futures.

It was not just ordinary villagers who were opposed to the dictates of projects of reform that linked underdevelopment to their own failure to overcome a backward nature. Disappointment with the failure of the mountain state to provide for and protect its *pahari* citizens had produced a new kind of politics whereby lack of development and opportunity was blamed on the growing presence of outsiders who not only expropriated *pahari* resources but also did not understand the significance of their distinct cultural imaginaries and practices. It was precisely out of this disappointment that the Gwal Sena emerged, an organization founded in 2009 by former members of the BJP who were frustrated by what they saw as the party's refusal to address the particular needs of *paharis*. The aim of the organization, its founders declared, was to address the historic injustices experienced by *paharis*, particularly the continued denial of their demands for control over *jal, jangal,* and *zameen,* one of the central aims of the mass movement in the 1990s. It was outsiders, they argued, who were profiting from the creation of a new state, not the *paharis* for whom it had been created. Unless the state stepped in to protect *pahari* lands, forests, water, and culture from outsiders, *paharis* would soon disappear.

FIGURE 4. "Gwal Sena's message: We will stop to take a breath only after working, with your support, to realize the Uttarakhand of people's dreams." Graffiti scrawled by Gwal Sena members on the side of a tank in Pokhri.

I first encountered the Gwal Sena where I encountered the PFA—in the battleground over sacrifice (Govindrajan 2014) and also in the growing storm over crop raiding and predation by wild animals such as monkeys, wild boar, and leopards. For them, as for the PFA, people's everyday relations with animals were at the heart of the messy business of defining what constituted *pahariness* in these complicated times. In chapter 5, I describe how the Gwal Sena joined villagers in insisting that rhesus monkeys who had been relocated from cities to provide relief to urban residents were key actors in a nefarious plot by outsiders to further expropriate *paharis* by making it difficult for them to sustain agriculture so that they would eventually have to sell their land. Like villagers, members of the Gwal Sena pointed to the aggressive behavior of the outsider monkeys as proof of their urban provenance and claimed that they were therefore unsuited to live with *paharis* in the mountains. In the chapter, I describe how villagers argued that the animal-rights activists who stepped in to protect these monkeys from culling or sterilization did not understand that there was a difference between living with *pahari* monkeys and having to share precious space and resources with monkeys from elsewhere who did not recognize the mores and obligations of the social world that they had newly entered. In other words, they insisted on recognizing the distinctiveness of different kinds of monkeys, a claim they extended to other animals as well.

Mohan, one of my neighbors, summed up this view rather pithily on a day when I had frustrated him with my repeated questions about what distinguished a *pahari* animal from a *desi* (plains) animal. "Look, are people not different from place to place? Are your friends in America like us?" When I shook my head, he proceeded. "So why should the animals be any different? You have come here to learn about *pahari* culture. To do so, you sit with us, eat with us, write notes about us in your notebook. If you want to learn about *pahari* animals, you will have to do the same with them. Sit with them, live with them, and you will understand why a *pahari* animal is different from a *desi* animal."

Animal Subjects

Mohan's exhortation brought to my mind the call to action in Eben Kirksey and Stefan Helmreich's (2010, 545) influential manifesto for multispecies ethnography, namely, the importance of treating creatures other than the human as ethnographic subjects with "legibly biographical and political lives." Over the last decade an important body of scholarship—drawing together science and technology studies, philosophy, history, politics, and literature—has focused on tracing the entanglements between humans and nonhumans and illuminating how, through these intimate encounters, they "become with" one another, to borrow Donna Haraway's (2008) now famous formulation.[19] Multispecies ethnographers are tracing the contours of multispecies worlds that are brought to life through the worlding labor of a multiplicity of beings, of which humans are only one kind. As Anna Tsing (2015, 23) reminds us, these multispecies worlds emerge when "gatherings become happenings, that is, greater than the sum of their parts." These entanglements make it ethically and experientially impossible to think of human lives—their histories and futures—in isolation any longer; indeed, "we" (as humans, although it is important to remember that many humans have always done so) must cultivate "forms of attention" that recognize the presence of all the critters who have become with us and shaped our becoming in turn (Van Dooren, Kirksey, and Münster 2016). The urgency of this work is driven by the growing recognition that it is no longer possibly to deny that in the "world beyond the human we sometimes find things we feel comfortable attributing only to ourselves" (Kohn 2013). These "things" that have, for so long, been seen as the exclusive preserve of humans include agency, subjectivity, intention, and self-consciousness—the hallmarks of the autonomous individual fetishized in Enlightenment philosophy.

Throughout this book, I draw inspiration from these recent inquiries and treat animals, human and nonhuman, as coparticipants in meaningful worlding. I am especially drawn to work that extends subjectivity to nonhuman animals and approaches them as persons whose inner lives and affective states are critically shaped by their experience of life in a world they inhabit in relation to a host of others. In other words, I adopt an approach that treats their reflexivity as a quality that is simultaneously intrinsic and relationally produced. On a related note, animals are also now acknowledged as crafters of stories about the world they inhabit and their relationship to it, a claim that I find compelling and transformative. In their wonderful article on urban penguin colonies, Thom Van Dooren and Deborah Bird Rose (2012) argue that nonhuman animals dwell in "storied worlds," that is, places about which they produce meaningful narratives through their actions. Narrative, they note, is no longer the sole preserve of humans; another pillar of human exceptionalism has crumbled. To write about multispecies worlds for them, then, is to write about the "storied experience" of both human *and* nonhuman persons and how they intersect and shape one another. The spirit of this multispecies storytelling animates the rest of this book.

In what follows, I explore how reflexive exchanges between *particular* humans and animals, facilitated through an embodied, touchy-feely, language of mutual recognition and response, were crucial to their coconstitution as subjects. In thinking of becoming as a process that is grounded in these bodily conversations, I draw once again on Donna Haraway, who describes such chatter as a "gestural, never literal, always implicit, corporeal invitation to risk copresence, to risk another level of communication" (Haraway 2008, 239). I emphasize *particular* above because, as the philosopher Vinciane Despret (2008) points out, animals who "talk" do not do so in the name of a "we" where the "we" represents an entire species "but in the name of a "we" constituted by the assemblage" of the animal(s) and the person(s) between whom conversation takes place. In other words, it is in the singular copresence of a particular set of individuals that one must locate these inquiries about narrative, subjectivity, and relationality.

It is to this end that the rest of the book lays emphasis on following the lives of *both* particular classes of animals—such as urban monkeys or mountain cows—and singular animals. To live up to the ethnographic promise implied in its title, the field of multispecies ethnography must focus on tracing the trajectories and outcomes of individual animal lives to see how they love, loathe, grieve, play, crave, and, indeed, relate. The difficulty is that in speaking of matter, the nonhuman, the other-than-human, or even the animal, we speak so generally that we lose sight of smallness and singularity—the goat who follows his beloved human everywhere, the monkey whose ear was bitten off in a bitter fight, the bear who builds a nest in the hollow of the same *deodar* tree every year. It is hard to grasp, no matter how imperfectly or incompletely, the intelligence of animals, their capacities for language, the richness of their emotional experience, or the meaning of their stories when one speaks of them only in the abstract, *animal*, or always as part of a collective, which is often the case in multispecies ethnography. At these scales, it is hard to imagine the tangible forms that relatedness might take between animals who encounter one another not just as part of a collective but also as "transcorporeal subjects"—individual bodies that are open to and always in the process of being substantially transformed by affective encounters with other bodies (Alaimo 2010).

What I want to emphasize throughout this book, then, is the importance of grounding theoretical insights about the capacities of animals in the lives of *actual, real* animals, both singular and collective. To paraphrase Hugh Raffles (2012), this means treating a particular animal as animal, the animal, and an animal, a process that "forces us to reimagine scalar relations."[20] To do so, I tell a "rush of stories" that move across these different scales—from entire species to particular classes to singular animals and back again (Tsing 2015; see also Sivaramakrishnan and Agrawal 2003). The scales themselves are inextricably connected. People made judgements about classes of animals based on their interaction with a singular animal. The intimate experience of the idiosyncratic behavior of a beloved *pahari* cow could be crucial to how people perceived and related to the entire class of mountain cows. Similarly, people's imaginings and experience of a particular collective of animals such as urban monkeys shaped how they responded and related to individuals of that class. The stories of multispecies relatedness I describe took shape through lively and transformative encounters across these intersecting, incommensurable scales.

I immersed myself in the lives of animals much as I immersed myself in the lives of people: by spending long periods of time with them; by closely observing their preferences and habits; and by learning how to "talk" to them through "messages exchanged in gestures," a subject I will return to at greater length in chapter 2 (Haraway 2008, 239). In other words, I practiced what Anna Tsing (2015, 37) calls "arts of noticing," immersive acts that engage sight, sound, smell, touch, and taste. Of course, some animals were easier to notice than others, both as species and as individuals. I came to be much better acquainted with goats, cows, monkeys, and dogs than with wild boar, bears, and leopards because they were more tolerant of my presence, not to mention less dangerous to hang around. Not all of them, however, appreciated my efforts to strike up a friendship. One adult female rhesus macaque, whom I recognized by the fact that her teats pointed inwards at one another, grew to resent my presence around her group so much that she would stare at me with an open mouth, a low-intensity threatening gesture, every time I walked past her. On a few occasions, when I did not learn my lesson, she would bare her teeth and then chase me for a few yards until I stood at a respectful distance. Binduli, my friend Manju's cow, would attempt to gore me any time she caught me alone. Manju's reassurance that Binduli was not like this with others and that it was only me she had seemed to take an instant dislike to, was not particularly comforting. My feelings were somewhat assuaged by the affection that Binduli's stallmate Rani showered on me, but, perhaps unsurprisingly, rejection by an ethnographic subject was no less wounding coming from an animal than it was from a human.

There were some animals that I came to know mostly or almost entirely through their material and immaterial traces–green bear scat, covered in loud, iridescent flies, lying in the middle of a mountain path; the half-eaten bovine carcass that a leopard had carefully dragged to the side of the road; the loud squeals and grunts of a particular group of wild boar who made regular, nocturnal visits to the fields directly below a house I lived in. These traces were visceral evocations of the animals who left them. Even if one caught only a fleeting glimpse of the leopard who lived with her cub on the rocky slopes looming over the village, her calls echoed through the village in the evenings and were a powerful sign of her presence in our midst. Importantly, these intimacies were no less meaningful because of their ephemeral nature. As Juno Parreñas (2016) argues, intimacy with some nonhuman animals can take unexpected forms. Intimacy, then, can be "about directly coming into contact with bodily effects and not necessarily actual bodies" (Parreñas 2016, 120). This helps to remind us, as Parreñas notes, that "the effects of bodies can stand in for actual bodies," creating intimacies that can be as powerful as those that emerge through physical contact between bodies (Parreñas 2016, 120). The forms of relatedness that I trace over the rest of the book emerge through both tangible and intangible, material and immaterial, proximate and distant encounters.

Through such encounters, I came to really appreciate how paying ethnographic attention—that is, sustained and careful attention—to what individual animals *do* over the course of their daily lives could yield valuable insight into their lives as empathetic, intentional, interpretive, and intelligent beings. The fact that these capacities are differentially distributed across species and easier to perceive in the case of some animals over others only strengthens the case for an ethnographic consideration of how agency, intention, and emotion actually manifest themselves in specific instances during the everyday course of social life. Consider, for example, the following stories. On a sunny day in 2014, a group of young rhesus monkeys were playing atop a low brick wall on the other side of which was a steep drop down into the valley. The wall had been built in that spot, right next to a *chai* shop, to prevent tourists and children from getting too close to the edge. One of the monkeys slipped and fell while bounding toward another over the narrow ledge and immediately disappeared from view. Some men peered over the wall and announced that the monkey was either dead or would soon die given the distance it had fallen. Imagine our surprise, then, when twenty minutes later, the monkey who had fallen was hauled up the slope by a couple of his companions. As we watched, the rescuers nudged, licked, and cradled their dazed relative for fifteen or twenty minutes until he was finally on his feet again. "*Wah*, what friendship," one man marveled, echoing my own thoughts. We all felt beyond a shadow of doubt that what we had witnessed was an intentional act of friendship, love, and empathy.

A few days later, I was powerfully reminded of this act of animal friendship upon witnessing another. My neighbors had decided to sell Sharmili, one of their two cows, because Gauri, the other, was pregnant, and they would need to build an expensive extension to their *goth* to house three cows. The trouble began within a few hours of Sharmili leaving with her new owners, a family in the next village. Gauri began to scream so loudly that her anguished bellows drew several of us to the shed. There we found Hema, my neighbor, who was trying to quiet Gauri by stroking her. Her tone was twisted with fear as she kept asking Gauri "ki haro?" (what happened?). After checking to make sure that Gauri hadn't been bitten by a snake, Hema realized that she was missing Sharmili. "Hopefully she'll get over it in a couple of days," she told us as we hurried away. Gauri's cries had chilled me to the bone, and I could tell others were disturbed by it too. A couple of the older women kept whispering "poor thing" to each other as they turned back to look at Gauri, whose body was visibly trembling as she poured her heart into each mournful roar. Her grief lasted three days before Hema decided that something needed to be done.

The next day, Sharmili returned home. When I went over to see how Gauri was doing now that her friend had returned, I found the two cows lying next to one another in the shed, with Gauri's body leaning heavily on Sharmili's. Sharmili was bathing Gauri's ear with her thick bristly tongue while Gauri's eyes closed with pleasure. They looked happy and at peace. Hema's father-in-law, who had accompanied me to the *goth*, told me that he had almost wept himself when Gauri was finally reunited with Sharmili. "She heard Sharmili call from the road above, and she got so excited that she tried to kick down the walls of the shed. She almost broke the rope [tethering her]. She has not left Sharmili alone for a minute since. They never used to eat from the same pile of grass earlier. But look at them now." Both cows were pulling strands of fresh grass from the large pile before them. Their bodies were still firmly pressed against one another. It was as if they needed constant, *bodily* confirmation of their reunion. Hema's father-in-law told me that they were planning to start work on the extension to the *goth* next week. When I said that it was worth it given how much their companionship meant to Gauri and Sharmili, he agreed. "Animals have a big *heart*," he told me. "They make deep relationships. We must respect those relationships."

There was no doubt in this man's mind that Gauri and Sharmili shared a rich and emotionally satisfying relationship that was crucial to their everyday sustenance; Gauri's sorrow was such that it might have killed her had Hema not decided to return the money from Sharmili's sale and bring her home. Indeed, all the humans I knew believed that nonhuman animals were intentional and subjective persons like themselves and that relationships with them could be successful only if they emerged from that recognition. As Frans de Waal (2016) writes in his recent book *Are We Smart Enough to Know How Smart Animals Are?*, for much of the last century science was overly cautious and skeptical about the intelligence of animals; anthropomorphism was, and still remains for many, an unpardonable sin. But the situation, as De Waal (2016, 4) gladly notes, is rapidly changing: "nothing is off limits anymore, not even the rationality that was once considered humanity's trademark." De Waal's (1997, 51) assertion that anthropodenial—a denial "of the human-life characteristics of animals, or the animal-like characteristics of ourselves"—is far more pernicious than anthropomorphism is a crucial intervention into ongoing debates about the stakes and contexts of human interactions with nonhuman animals. To move out of anthropodenial, then, will require an extension of the creative imagination—a "sensory attentiveness to qualitative singularities . . . and [the willingness] to admit a 'playful element' into one's thinking"—as we learn about the nature of animal lives (Bennett 2010, 15).

The stakes of such openness cannot be overstated. In his searing exploration of the industrialized slaughter of livestock animals, Timothy Pachirat (2011, 40) points out that it is the erasure of an animal's individuality that allows her or him to be produced as "raw material . . . one thing identical to the thing next to it, which is identical to the thousands of things next to it, all ready to be fabricated into a series of meat products." If mass man-made slaughter is to be questioned and challenged, we *must* embrace a critical anthropomorphism that refuses this erasure by insisting on the creativity, intelligence, and emotional capacity of individual animals, a proposition that I will return to in the epilogue.

Let me repeat a cautionary note I have sounded elsewhere in this chapter. Throughout this book, I argue that relatedness does not overcome or erase ontological differences between human and nonhuman animals. I do not wish to suggest that in "becoming with" one another, humans and animals become *one*. The elision of difference between diverse beings, even if motivated by a desire to challenge human exceptionalism and the untenable boundary between nature and culture, often ends up reinforcing the domination of the human by drawing the other-than-human into the ambit of human experience while creating a false veneer of equality. Given these dangers, it is important for us to remember that the other-than-human, as Stuart McLean (2016) puts it, remains an "intimate stranger," a "force that can never be exhaustively encompassed by human intentionalities and understandings."

I therefore propose that relatedness must always be understood as constituting a partial connection between beings who come to their relationship as unpredictable, unknowable, and unequal entities. The forms of relatedness I track over the following chapters emerge from a recognition that the proclivities, emotions, and desires of different animals are dissimilar and can never be fully reconciled even as they are constituted relationally. Unlike other multispecies ethnographies that tend to focus on human relationships with a single species—whether a primate, an insect, or a canine—this book's exploration of the lives of a range of different animals enables it to more easily track the variations between and within multiple species. As the anthropologist Barbara King (2013, 8) notes, the "great lesson of twentieth-century animal behavior research was that there is no one way to be chimpanzee or goat or chicken, just as there is no one way to be human." Relatedness, then, gestures to one's existence in a world where the overlapping of different lives and fates means that difference had to be constantly and imperfectly negotiated through shifting turns to love, care, neglect, avoidance, and violence.

This recognition of difference has implications for my own understanding of the animals I encountered during my fieldwork in Kumaon. Given that my access to them was partial and fragmented, the lives of the animals I studied are certainty not exhausted by the scope of my knowledge about them, the knowledge I offer to you, the reader, in what follows. Given this difference, then, how should one characterize an ethnography that counts nonhuman animals among its subjects? One possible answer comes from the philosopher Timothy Morton (2011, 166) who puts the matter thus: "objects withdraw such that other objects never adequately capture but only (inadequately) 'translate' them." The idea that one can only produce an incomplete and inadequate translation of how one's subjects feel, think, and act is, of course, familiar to the ethnographer. Ethnographies are partial renditions of an experience that often exceeds translation. The work of Marisol de la Cadena offers another path to think with difference. In *Earth Beings*, Cadena (2015, 32) notes that there are occasions when "no translation would be capacious enough to allow [one] to *know* certain practices." Yet if one thinks of translation as Cadena does—in terms of communication that is not hungry for commensuration— there is still room for curiosity, the bedrock of ethnographic practice. Differences, she writes, "also connect. . . . The idea that differences can connect rather than separate . . . suggest[s] an anthropological practice that acknowledges the difference between the world of the anthropologist and the world of others, and dwells on such differences because they are the connections that enable ethnographic conversations." While the translation of difference I offer you in the pages to come is certainly incomplete, I hope its deficiencies are attenuated by the deep curiosity and constant sense of wonder that fuels it.

A Note on Place and Method

In 2010, I began fieldwork in a cluster of Kumaoni villages anchored by the town of Mukteshwar. Mukteshwar's fame as a beautiful and salubrious hill station goes back to colonial times. In his account of hunting a man-eating tigress that killed several villagers in this area, Jim Corbett, the much-feted colonial hunter, wrote that "people who have lived at Muktesar claim that it is the most beautiful spot in Kumaon, and that its climate has no equal." It was the climate and forests of the region that led colonial administrators to establish the IVRI in this town. Even today, the IVRI is one of the largest local employers, its presence made hypervisible in the cluster of colonial bungalows, kraals, and laboratories that dot the mountainside and that can be seen from miles away. The region is also home to several important NGOs that have invested heavily in rural development projects over the last several decades. When I started my fieldwork, I was loosely affiliated with Chirag, a rural development organization. For the first couple of months of fieldwork, I lived in Chirag's guest house. After traveling to several villages and living with different families there, I decided to base myself primarily in the village of Pokhri.

Pokhri attracted my attention for several reasons. First, it was a village that was home to a number of different *jatis* or castes, both high and low.[21] Only some *jatis* were original to the village, which itself had been relocated in the early twentieth century from within the forest managed by the IVRI to its fringes; others had moved there from ancestral villages in what are now

FIGURE 5. The village of Pokhri with the IVRI forest in the background.

Nainital and Almora districts in search of employment with the IVRI. Second, several estates that had belonged to colonial officials and that were then handed over to their creditors—the Sahs—at the time of India's independence were located in Pokhri. Much of this land had been sold by the Sahs either to individuals from Delhi, Mumbai, and other cities in India's plains or to real-estate developers who wanted this land to build second homes and hotels. The "village," as I said earlier, was a mix of villagers, urban elites who often thought of themselves as villagers, and NGO workers. This was by no means a "remote" place; indeed, many tourists often complained that it was less isolated than they expected. As such, my location in the fluid and complicated world of Pokhri allowed me to understand how people's relationships with the livestock and companion animals they raised as well as with the wild animals who moved in and out of this patchwork landscape of field and forest were fundamentally shaped at the intersection of different ideologies and structures of power and modernity.

In addition to Pokhri, I lived for shorter periods of time in two other villages—Loshi and Danka. Loshi was a village that was particularly plagued by the depredations of wild boar and monkeys. Almost every time I mentioned that I was interested in studying "human-wildlife conflict," people would point me toward Loshi and its surrounding villages as an example of the deleterious effects of living in proximity with wild animals. I first traveled to Danka, the third village, with a woman from Pokhri whose *mait* (maternal

FIGURE 6. The village of Loshi.

home) it was. Danka was widely acknowledged as a repository of Kumaoni culture; I was told by multiple people in Pokhri and Loshi that I should go to Danka if I wanted to learn about Kumaoni customs and folklore. In addition to these three villages, I traveled to many other villages for weddings, festivals, and intimate household rituals in the company of people from Pokhri, Loshi, and Danka. Quite often, these travels were initiated when young women I knew well asked me to accompany them to their *mait* so that I could meet their maternal families. In the pages that follow, I offer stories from my time in these different villages that, in the interests of maintaining the anonymity of those whose lives and thoughts I track, I do not always specify by name.

It would be remiss of me if I did not mention how much of my time in these villages was spent in the company of women, not just because I am a woman myself but also because they were largely responsible for the care of domestic animals and for a variety of agrarian tasks that brought them into contact with wild animals. Chapter 6, which delves into a genre of stories about black bears who abduct and have sex with women, draws much of its material from the women I came to know well through the extended time we spent together at the grazing grounds or in their kitchens or during festivals. It was because they let me into their world as a friend, as a daughter, as a sister, and as a sister-in-law that I came to see clearly how human relatedness to animals was inflected by gender. In that chapter, I argue that the transgressive desires celebrated by women in their accounts of these sexual encounters call into question not just patriarchal but also anthropocentric hierarchies in which the boundary between humans and nonhumans is inked on the terrain of desire, a troublesome quality for women, in particular, to possess. I argue that what makes these stories so compelling is the fact that women come to know and relate to some animals differently from men; this difference, I suggest, can be attributed to the gendered nature of the labor involved in creating and sustaining interspecies relationships. Both in this chapter and elsewhere in the book, the account of relatedness I provide thus attends not only to how interspecies connection can take different forms depending on the kind of nonhuman animal that is engaged but also to how understandings and experiences of what it means to live a life in relation to another shift across different kinds of humans depending on their caste, class, and gender, among other things.

Before we move on to the rest of the book, allow me to take you back briefly to a hot Sunday morning in September 2010 when my friend Kusum asked me to take her goats out to graze for the day. Her older daughter, Rajni, who usually took the goats to graze on her days off from school, was unwell, and Kusum herself was busy with cooking an elaborate lunch for her husband's relatives who were visiting from the town of Haldwani. Kusum handed me a stick and some *rotis* as I prepared to set off for the day. I was not to go far, she said, just around the stream that flowed through their land; there was plenty of green grass there, and the goats could fill their stomachs without my having to run around after them. The morning was an utter disaster. The goats who, it seemed, had grasped my lack of expertise soon scattered across Kusum's fields, with one crossing the stream into a neighbor's property, something Kusum had warned me not to let happen. By the time I retrieved the drifter, the others had disappeared. It was with great difficulty that I rounded up the goats after a few hours and returned to Kusum's house. I was almost in tears when I recounted the events of the day to her. She tried valiantly not to laugh when I told her how one of the older goats, the one who had crossed the stream despite my numerous attempts to deter him, had reared up repeatedly on his hind legs and butted me while I looked for something to tie around his neck so that I could lead him back. He had bleated his displeasure all the way home, drawing the attention of several villagers who laughed when they saw how I was struggling to make him follow me. "It's no matter," Kusum said comfortingly as I apologized for my failure. "They don't know you, which is why they decided to mess with you [*panga lena chaha*]. It will take some time. But once you understand them and they understand you, a relationship will be formed. Then you see, they will do whatever you tell them to." I suspect that Kusum exaggerated the eventual willingness of goats to follow instructions in order to comfort me; after all, she spent most of her time grumbling about how difficult it was to manage her herd. However, her sage advice proved to be true. As I spent more time around her goats, I came to know them better and finally started to understand how to make them respond to me, even if the results were mixed. My old adversary, for instance, never missed a chance to butt me when I wasn't on my guard. A few months later when I was able to take the goats out without much incident, Kusum was appreciative. "See," she said, "I told you the relationship would be made. It's hard work, that's all." She paused for a few seconds before warning me that I would have to spend more time with the goats if I wanted their recognition (*pehchan*) of me to last. She was only half joking. Kusum's observation about the hard work required to sustain and negotiate a fraught relationship has stayed with me since that day so early in my fieldwork, and it is in the spirit of the fragile promise of mutual recognition nested in relatedness she evoked that I offer you the rest of this book.

The Goat Who Died for Family
Sacrificial Ethics and Kinship

What is the nature of sacrificial connection between the one who sacrifices, the one who is sacrificed, and the one who accepts the sacrifice? Does sacrifice leave an imprint on everyday relationships that extends beyond the moment of ritual killing? I was brought to a consideration of these questions after a visit to the Kalika temple in Gangolihaat during the Navratri season in 2011.[1] On Ashtami, the eighth day of the Navratri (lit. nine nights, a festival commemorating the victory of the goddess Durga over the buffalo demon), I reached the temple in the company of a family who had brought with them eight goats— all male, since only bucks could be sacrificed—to be sacrificed to the *devi*. Neema Bhandari, the formidable fifty-three-year-old woman who invited me to accompany the party, had organized the *puja* (ritual offering). Years ago, Neema had prayed to Ma Kalika, a powerful Shakta goddess and the patron deity of the Kumaon Regiment, to bless her husband, Puran, with a successful military career.[2] When he eventually retired with the rank of honorary captain, an achievement that no one in their family or village could match, it was clear to Neema that the *devi* had responded to her fervent prayers. It was now Neema's turn to make good on her promise—an *athwar*, the sacrifice of eight animals.

It had taken Neema some time to save money for the *puja*. Two years earlier, she had bought eight kids from her sisters, whose goats had recently given birth, and dedicated them to the goddess. She had taken care of the kids herself, she told me proudly. No one else in the family had been allowed to care for them. Then there was the expense of the feast that would follow the *puja*. When sufficient funds were finally in hand, Neema decided that the *puja* would be offered during Navratri.

FIGURE 7. The family enters the courtyard with the goats.

In the wee hours of the morning, a cavalcade of cars, Tempos, and motorcycles had arrived at their village to transport people to the temple of Ma Kalika. After being anointed with *chandan* (sandalwood) and *pithiya* (vermillion) by the village priest, the goats were crammed into the back of a cream-colored Tempo. Two of Neema's teenage nephews clambered into the stuffed space with the animals and held onto the iron frame for balance as the Tempo started with a smoky sputter. Following on a motorcycle, I watched the boys jostling for space with the goats as they danced to the beat of the drum that somebody was playing in one of the cars that made up the procession. After arriving at the Kalika temple a few hours later, the family rang the brass bell at the red archway in the entryway and then took the cement path that circled the shrine instead of going directly inside. We completed the customary circumambulation of the shrine and its grounds, stopping every now and then to cheer *Jai Ma Kalika* (Victory to Ma Kalika). Several members of the family, including Neema's daughter who was making her first visit to her *mait* after marriage, were possessed by their *kul dyavtas* (household deities).[3] The other women ululated approvingly, delighted that the family deities had appeared on this important occasion.

Pilgrims thronged the inner courtyard of the temple when we walked in. There, the family, with their goats in tow, were greeted and blessed by the elderly *pujari* who tended to the main shrine. Kneeling with some difficulty on the cool marble stones donated by the Indian army and taking special care to avoid the pellets of goat dung that studded the floor like raisins, the priest applied more sandalwood and vermillion to their foreheads. He then touched a drop of water to the mouth of each goat, inducting them into the Bhandari family *gotra*.[4] Puran, and later the priest, explained to me that by being inducted into the family *gotra*, the goat had taken a *samkalp*, a vow to complete a particular religious task. The goat was, in essence, taking a vow to sacrifice himself to a deity on behalf of the family of which he was now part.

A mixture of uncooked rice and water was then sprinkled on the goats' backs. The family held their breath until each goat shook his body, a movement that Neema described as *jharr*. This *jharr* was read as a sign that the goat had consented to his own death and that the deity was pleased with and had accepted the sacrifice. If a goat did not shake immediately, more rice and water was scattered on his back until he did. Once all the goats had shaken their consent, they were whisked off to an open air shed below the courtyard that had been set up as a staging ground for the sacrifices. I watched as the goats stiffened when they inhaled the tang of blood that rose up from the wine red ground. One of them, who had a snowy-white coat with beautiful, long ears the color of dark honey, moaned repeatedly, his tone increasingly desperate. I was sick with anxiety as I heard his fear. Neema leaned toward me and whispered, "he's calling for me. That's how he would cry when I locked them up in the shed at the end of the day." Her voice broke as she turned her face away.

The sacrificer, a tall man resplendent in blue and gold silk with telling splatters of blood all over his face and clothes, made quick work of the goats, chopping the head off each one with a surprising ease. As each headless body was in the throes of a final, fierce spasm, the head was handed to an assistant to keep aside. The head and one hind leg of each goat was the sacrificer's to keep, and the rest of the meat was given to the family. Standing next to Neema, tall and graceful in a yellow chiffon *saree* that a niece from Delhi had given her, I felt each time she winced when one of the goats was beheaded. She was distraught when the goat with the honey-colored ears was killed, and tears started to roll down her cheeks. Meanwhile, the young men in the family hoisted bloodstained gunny sacks stuffed with meat onto their backs, and walked up the steep path to where the cars were parked. The sacks were loaded into a van, ready to be transported back to the village where the meat would be served at the feast that evening. A small portion of the meat was cooked right there and served to members of the family as *bhog* or *prasad* (food exchanges with divinities). The distinctive scent of singed hair and flesh hung heavy in the air and lingered on people's clothes and in their hair.

Through all this, Girish, Neema's nephew, who was visiting from the United States, had stood conspicuously apart from the rest of the family. After

FIGURE 8. The sacrificer kept a hind leg from each goat as his due.

we returned home after the *puja*, Neema berated him for refusing to eat even the *prasad*, an act that she termed an insult to the *devi*. When she demanded to know why he had behaved in that manner, Girish responded that he was struck dumb by how "backward" his family still was. Sacrifice, he said bluntly to the assembled family, was a "barbaric" practice. Why couldn't the family have offered the *devi* flowers or coconuts? Insisting that deities would be as pleased with vegetal offerings as with a goat, he concluded that it was people who really wanted to sacrifice animals so that they could eat meat and show off their wealth to the rest of the village. "Do you really think our *dyavtas* eat meat?" he asked in a voice dripping with scorn.

In response, Neema told us the story of how the practice of animal sacrifice had started in the mountains. "In the old days," she began, "the goddess would be satisfied only with blood."

> Each household that took turns to serve in her temple would offer her their oldest son. Once it was the turn of an old widow who had to offer her only child as sacrifice. She couldn't bear the thought and decided to appeal to the *devi* for mercy. The next morning, she waited by the *naula* [water tank] where the *devi* bathed every morning. When the *devi*'s retinue arrived, the old woman fell to her knees and begged for the life of her son. The *devi* asked, "What will I get in return?" So the old woman promised to sacrifice the animals that she had raised just like her own children. The *devi* relented only because of that promise. . . . If we don't offer the *devi* a *puja* [sacrifice] . . . then we will be back to the terrible days of *narbali* [human sacrifice]. This is why [all this talk of]

offering coconuts and flowers as *puja* is nonsense . . . They are not precious.
There's no loss when you give *devi-dyavta* coconuts. But giving an animal is
like giving a human. A life is given in place of a life.

An uncle-in-law of Neema's entered the fray and pointed out that flowers and
coconuts would not satisfy the *dyavta*'s desire for *maans* (flesh) and blood.
Besides, he said, goats were also *bhakts* (devotees), and "gods expect the life of
a *bhakt* because it tells them how much *bhakti* [devotion] they can command.
A coconut isn't a devotee," he said; "an animal is."

While everyone else in the room seemed impressed with the simple clarity of this statement, Girish was still unsatisfied. He repeated that people in
the mountains were sacrificing innocent animals in their own interest and
declared that it was a sin to "murder" a living being in this manner. Neema
snapped at him. "Do you know how much labor it takes to raise animals?" she
cried, her voice shrill.

> I work day and night to take care of these animals. My legs are tired from walk-
> ing with them all day, my arms are still scarred from when I entered a thicket
> of thorns to rescue one of the young ones. I have cared more for them than my
> children . . . Taking care of animals is an everyday ritual. But you see only the
> ritual of sacrifice and then say that we don't really love our animals. It pains
> me every time I see one [of them] die. I feel such *mamta* [maternal love] for
> them . . . I think of them even after they are dead, I cry for them when I open
> the door to the shed and they are gone . . . They repay the debt of my *mamta*
> [by dying for our family].

Girish's response was a sarcastic snort of laughter; he was visibly agitated.
This talk of *mamta*, he said, was meaningless. After all, *mamta* didn't prevent
people from sending these animals to their death in place of their sons. His
aunt's eyes shone with angry tears at this dismissal of her claim to the maternal
affect that emerged through acts of gendered labor. "You may not concede my
mamta," she said, finally, "but the *devi* can see it. She knows it was a true [*sach*]
sacrifice for me . . . like watching a child die."

Girish is not alone in his objections to animal sacrifice. The practice of
animal sacrifice is regarded with hostility by animal-rights activists, Hindu reform organizations, and a number of *pahari* youth. Simmering tensions about
sacrifice boiled over in 2010 when the PFA filed a Public Interest Litigation in
the Uttarakhand High Court calling for a ban on the practice on the grounds
that it not only entailed cruelty against animals but was also a false superstition that violated the Hindu ideal of *ahimsa* (nonviolence). In its ruling on the
writ petition in 2011, the High Court acknowledged that there was a long tradition of sacrificing animals to local deities but declared animals could no longer

be sacrificed for *explicitly* religious reasons. The judgment stated that "the person sacrificing an animal can only sacrifice the same, *not for the purpose of appeasing the Gods*, as he believes, but only for the purpose of arranging food for mankind."[5] Significantly, it went on to say that animals slaughtered for food could be killed in a manner dictated by any particular religion, but they could not be sacrificed to deities in a temple. Slaughter for food, the court stipulated, would now take place only in slaughterhouses recognized by municipal or other local authorities, *except* in rural areas, where this rule was more difficult to enforce. With this ruling, the court moved sacrifice out of the realm of ritual practice and into the secular domain of food provisioning. Sacrifice would be acceptable only if stripped of its devotional character; the deities to whom the gift was originally offered were rendered incidental and the sacrificial animal transformed from a ritual substitute for a human victim to a mere source of food.

Neema's argument with Girish reveals much about what is at stake in the vigorous and often angry debates that were taking place across the region on the meaning of sacrifice. For Girish and others like him, the association between goats and children was limited, even hypocritical, in that people did not offer their children as sacrifice even as thousands of goats were sacrificed every year. The claim that a sacrifice becomes real only by offering a being whose death causes pain to the sacrificer was regarded by many opponents of the practice as opportunistic, a thin veneer for what they saw as senseless cruelty and violence. The ethical questions opponents of sacrifice raise about human responsibility for the extinguishing of animal life cannot be evaded. Indeed, those who continued to offer animal sacrifice were, more often than not, troubled by its violence, and they attempted to navigate the ethical dilemmas it raises in myriad, complex ways. As such, Girish's assertion that Neema's experience of loss and grief in the face of death was simulated forecloses any engagement with the complex entanglements of care, kinship, and violence that characterize these sacrificial relationships.

In this chapter, I argue that perspectives like Neema's offer compelling responses to the two questions that I posed at the beginning of this chapter. On the nature of sacrificial connection, her point of view is that sacrifice is made meaningful and authentic, and by extension acceptable to the gods, by shared bonds of kinship—what she experienced as *mamta*—between the person who offers the sacrifice and the animal that is sacrificed. Following her, I explore the idea that sacrificial relationships are marked by practices of care, attention, and reciprocity that emerge through everyday, gendered forms of labor involved in raising animals who are eventually sacrificed. When Neema described the scratches on her body received from crawling into thorny thickets to rescue a goat, it was this intimate, routinized, arduous, and affective labor that she was invoking. This daily entanglement of lives creates bodies that are open to being affected by one another, and it is this porosity that allows different beings to come together in relationships of proximate intimacy. The death of an animal with whom people feel this embodied kinship creates a sense of loss and grief that is essential to making sacrifice *truly* a sacrifice. This, I argue, is the nature of sacrificial connection.

Neema's response to her nephew Girish offers an interesting lens through which to consider my second question about whether ritual sacrifice has an influence on everyday life beyond the sacrificial moment. When Neema said that she mourned the death of her goats, she was not posturing but offered insight into the fact that the spectacular nature of sacrificial death forces people to engage with it long after it is over. I argue that this engagement takes the form of ethical acts and aspirations that emerge in the ordinary unfolding of everyday life even as they respond to an out of the ordinary event. These gestures are often small and fleeting, like Neema's tears on the occasion of her beloved goat's killing. Small and perhaps even insufficient though they may be, they open up the possibility for ethical kinship and love in the interstices of violence.

Sacrifice as Relatedness

Ritual sacrifice (*puja* or *bali*) remains an important part of the propitiation of a number of local *devi-dyavta*, ghosts, and unhappy spirits in Uttarakhand.[6] Not all *devi-dyavtas* ate sacrifice, but even "vegetarian" deities such as Golu *dyavta* could sometimes demand it for their guardian deities. *Khushi pujas*, like Neema's *athwar*, were offered in gratitude for acts of beneficence by *dyavtas*, whether the gift of children, jobs, or promotions. At the heart of these rituals is the belief that development could be successfully pursued only through the intervention of *dyavtas* who reward piety and hard work and act "as the ultimate trustees of terrestrial well-being" (Pandian 2008, 172). Devotees, in turn, must show their appreciation for these acts of divine favor. The regular performance of sacrificial rituals affirmed that *dyavtas* and *bhakts* existed *in relation* to one another; that devotion and care would be rewarded with economic, moral, and familial prosperity. These *pujas* were usually offered both at large regional temples and humbler village shrines constructed out of stone and mud. Out-married daughters of a village, for instance, were expected to offer a sacrifice to some of the *isht* (literally, "beloved" or "chosen"; a deity with whom devotees share an affective relationship) and *kul* (family and lineage) *dyavtas* of their natal village upon the birth of their first child.[7]

If such sacrifices were not given voluntarily (and quickly), *devi-dyavtas* often became angry (*bigad jaana*) and then had to be propitiated through a lengthy series of rituals intended to turn malevolence back into benevolence.

Anthropological literature on sacrifice understands the practice as a set of ritual narratives and actions that variously establish forms of transaction with the gods (Tylor 1871; Smith [1889] 1927); displace the sins or pollution of the sacrificer or societal violence more generally onto a scapegoat (Evans-Pritchard 1954, 1956; Girard 1979); link the sacred and profane through the intercession of a sacrificial victim (Hubert and Mauss 1964); or "ransom" the life of an original human victim from death (Dhavamony 1973; Smith and Doniger 1989).[8] As Marcel Mauss and Henri Hubert pointed out in their remarkable analysis of sacrifice, these motivations are not always mutually exclusive. Veena Das (1983, 445) offers a crucial corrective to the universalizing claims of these theories through her observation that attempts to outline "universal features of the sacrificial process draw rather heavily from assumptions about man, society and God in Semitic traditions." This is a problem that, as Talal Asad (1993) notes, extends to some of the foundational categories in the anthropology of religion as a whole. However, what is common to anthropological analyses of Christian and non-Christian forms of sacrifice is the agreement that the principle of substitution—the use of a surrogate victim for an original other—is crucial to the practice.[9] It is this "sleight of hand," a game of displacement and replacement, Wendy Doniger and Brian Smith argue, that distinguishes sacrifice from other forms of death. What makes this substitution possible, they suggest, is the "theoretical identification of the sacrificer and the victim" (Smith and Doniger 1989, 99).

For the most part, this body of work emphasizes the symbolic significance of the sacrificial victim to the ritual; little or no attention is paid to how its symbolization might be tied to its distinctive materiality. Evans-Pritchard (1953, 185), though he was otherwise attentive to the ways in which material entanglements between humans and oxen were important in the sacrificial process, argued that sacrificial identification was only with the "idea" of animals.[10] In a somewhat different vein but keeping with the idea that the sacrificial victim should be apprehended largely in symbolic terms, Smith and Doniger (1989, 195) suggest that the primary function of the victim is to serve as a "symbol for a pair of opposites: the sacred and the profane, god and the human, recipient and giver." Further, "because it is a symbol, it is not critical to the efficacy of the sacrificial process itself who or what is selected to act as the symbol, the victim" (Smith and Doniger 1989, 195). One reason for this scholarly emphasis on semiotics at the expense of the "who or what" of the sacrificial body might be the reliance on sacrificial myths and texts that, especially in the Vedic context of South Asia, as Wendy Doniger (2009, 153) notes, "treat substitution not as a historical question but as one relating to the nature of ritual symbolism, explaining how it is that plants or mantras stand for animals, and animals for humans, in the sacrifice."

The symbolic systems of classification that underlay sacrificial substitution struck Levi-Strauss (1966, 227–28) as "wanting in good sense" because they "adopt a conception of the natural series which is false from the objective point of view for . . . it represents it as continuous."[11] Levi-Strauss's characterization of the nature of substitution in sacrifice has been critiqued by several scholars.[12] For instance, Veena Das (1983, 456–57) points out that the natural world is *not*, in fact, represented as continuous in Vedic sacrificial systems.[13] However, Levi-Strauss's misgivings about the logic that informs sacrificial substitution do raise interesting and important questions. When we recall that Vedic texts included "man" among the five domestic animals who could be sacrificed (Doniger 2000, 2009; Das 1983), the questions that confront us are these: what properties allow an animal to symbolically substitute for humans in sacrifices where people were originally the prescribed victims? More specifically, if we return to Neema's story about how animal sacrifice replaced the original practice of human sacrifice in the mountains, how do we make sense of the *devi's* acceptance of an animal in place of a human? How can the sacrificial victim be understood as more than just symbol? How do we recover its history as a fleshy being whose semiotic qualities are embodied, and "always entangled . . . with material processes" (Kohn 2007, 5)?

In response to these questions, I offer the idea that systems of symbolic representation become meaningful when they are grounded in lived material relations, a theme that I will return to in the context of cow protection in chapter 3. To be more specific, the substitution of goat for human in sacrificial rituals in Uttarakhand makes "good sense" when we recognize that the production of sacrificial symbolism occurs in a world of interspecies reciprocity and relationality. Let me return to Neema's account of why the *devi* accepted animal sacrifice in place of an original human sacrifice. It was the promise, Neema said, to substitute a victim who would be raised with as much love and care as a child that caused the *devi* to relent. Neema's grief at the death of her goats bore testament to the depth of her *mamta*, an affective kinship that gave meaning to the substitution of goat for human in the sacrifice offered to the *devi*. A coconut, at least in Uttarakhand, is not imbricated in everyday relationships of care and maintenance and, indeed, love. As such, it is not symbolically significant. Coconuts, like vegetables, "remain stubbornly other"

(Coetzee et al 1999, 99), insignificant in comparison to human and nonhuman animals. Lucinda Ramberg's (2014) observation that one does not make offerings to the gods that are not valuable is apposite here.

Taking seriously the experience of Neema and others like her, I am therefore suggesting that the symbolic identification between human and animal that undergirds sacrificial substitution in Uttarakhand is made possible by a shared sense of kinship fostered largely through the embodied experience of being entangled in intimate relations of care and mutual subjection. I follow Sara Ahmed (2010, 22) in understanding this affective kinship as emergent in "the messiness of the experiential, the unfolding of bodies into worlds . . . [the process of being] touched by what comes near." The idea that sacrificial animals share a bond of kinship with those who sacrifice them is not new to anthropological scholarship on sacrifice. In his work on early Semitic sacrifice, Robertson Smith ([1889] 1927) argued that the original sacrificial animal was akin to a totemic animal believed to be holy by those who sacrificed it.[14] Human worshippers would, through the communal consumption of sacrificial meat, not only emphasize their blood relationship to the animal and the god but also reaffirm their blood bonds with one another. Freud (1918, 225), summarizing Smith's observations, writes that in prepastoral societies "the sacrificial animal was treated like one of kin; *the sacrificing community, its god, and the sacrificial animal were of the same blood*, and the members of a clan." In *Elementary Forms of Religion*, Emile Durkheim (2001, 192) captures something of the essence of this kinship when he notes that the aboriginal Gewwe Gal of New South Wales believed "that each person has within him an affinity for the spirit of some bird, beast or reptile. It is not that the individual is thought to be descended from that animal, but that a kinship is thought to exist between the spirit that animates the man and the spirit of the animal."

Indeed, as Kim TallBear (2011) points out, such expressions of human kinship and community with the more-than-human world abound in "indigenous or aboriginal voices." In his work on the peoples of the Kluane First Nation, for instance, Paul Nadasdy (2007, 27) argues that they "see themselves as embedded in a web of reciprocal relations with the animals on whom they depend. By accepting the gifts animals make of their own bodies, hunters incur a spiritual debt that they must repay through the observance of a whole series of different ritual attitudes and practices." I agree with Nadasdy about the need to remain open to the possibility that such declarations of kinship and reciprocity are potential truths, an argument that you might recall from the introduction.[15] When Neema and other *pahari* villagers articulate the importance of kinship, reciprocity, and debt in their relationships with animals whom they sacrifice, they ground these claims in the everyday acts of labor they perform in caring for these animals. It is this painstaking labor, as intense and prolonged as that entailed in rearing children, that is at the heart of the idea that because animals are like children, they can substitute for human children in sacrifice.

Labor and Love: The Stuff of Kinship

One summer day in 2011, I accompanied my friend Kusum and her herd of ten goats down a narrow mud path lined on either side with stinging nettles and fruit trees. We were on our way to the grazing grounds. One little white goat with brown spots had climbed up a small apricot tree. Ripe fruit fell to the ground as she pulled hard at the leaves. Before Kusum could fling a stone at her, an elderly farmer appeared on the other side of the low stone wall that bordered the path and proceeded to scold Kusum loudly for not controlling her animals. The week before this unexpected meeting, Kusum's goats had ravaged a wheat field belonging to the farmer, whose name was Narinder Bisht. Kusum's younger children had been with the goats that day but had fallen asleep while watching them. The goats promptly wandered into nearby fields and set to work, where they were caught in the act by Bisht. He tied them to a post and sat next to them, waiting to scold the children when they appeared to collect their goats. Unfortunately for him, he, too, made the mistake of falling asleep, whereupon the children, who had been watching him from a distance, quietly untied the goats and took them home, flushed with triumph at having escaped a scolding. Having caught Kusum now, Bisht let loose about her "thieving children" and "poisonous goats."

That evening, Kusum told her husband what had transpired. "It's these goats," she said at the end, "always the root of conflict. *Dusht* [wicked] animals." "Goats and children," her husband grumbled, casting an eye on their two girls, who were watching television instead of doing their homework. "They're both so troublesome. What can you do?"

Like Kusum's husband, other villagers often drew comparisons between raising children and raising goats. They emphasized similarities between the two especially when talking about how exasperating goats are; they noted their willfulness and tendency to run away when no one was watching them. "You have to scold both children and goats," a woman told me once, laughing as I struggled ineffectually to coax a goat away from a patch of green garlic that he was consuming with alarming speed. "If you had children or goats, you would know this. They don't understand anything without a beating." She threw a stone at the goat, and he moved off lazily as if to prove her point.

The other sense in which raising goats was similar to raising children was in the amount of work involved in both tasks. This is especially true as livestock rearing has become an increasingly popular element of rural livelihoods. Given the unreliability of agriculture in this terrain, the state has made sustained efforts over the last decade to popularize animal husbandry, especially goat keeping, as a livelihood strategy. As in the case of other farm-improvement movements across the world, these schemes are shaped by gender ideologies that represent farm labor and especially livestock keeping as feminized, and are therefore often targeted at women.

Women are expected to perform the bulk of the grueling labor involved in caring for livestock animals.[16] A large portion of every single day is spent in taking the animals out to graze. The poor quality of grazing grounds means that women have to work hard to feed their animals. During the monsoon, women stagger home in the evenings, backs bent under the enormous load of piles of green grass that they spend all day cutting. No matter the time of day in this season, the question "Gha katan chha? [cutting grass?]" is answered by "Hoy, gha katnu [yes, cutting grass]." Once the rains have passed, in the months when the heat from the sun is said to be strong enough to "dry a rhinoceros's hide," women focus on cutting and drying grass for the winter ahead. The arduous work of cutting grass, which many women blame for the chronic back and neck aches they experience, is supplemented by the equally strenuous year-round labor of making trips to the forest and climbing oak, *kaaphal*, and *pangar* trees to cut leaves that will be fed to livestock along with the dried grass. There is also the task of collecting fallen leaves and pine needles to serve as bedding for animals in the cold winter months. The daily routine of clearing dung from the shed, sometimes twice a day in the monsoon season, when the rain is too heavy to take animals out to graze, rounds out the long list of chores associated with raising animals. Little wonder, then, that women often speak of the work of raising animals as destroying their bodies.[17] Indeed, the performance of labor was a crucial element of women's subjectivity in the mountains. "You know what the identity of a *pahari* woman is?" a young woman asked me once. "Hard work . . . it starts from childhood itself. Women in the plains don't do this kind of work. But here a forty-year-old woman looks much older. Hard labor is our identity."

On the whole, raising goats requires far less labor than keeping cattle. Goats, often classed as microlivestock, are sustained mostly by open grazing and the occasional treat of freshly cut sweet-smelling grass. Since goat's milk is not consumed in the region, their nutrition is not a matter of particular concern. However, women often point out that goats require as much if not more patience, stamina, and hard work than cows because they are mischievous

FIGURE 9. Women must perform grueling labor in order to care for their animals. After the monsoons, they spend the better part of every day cutting grass for fodder.

and inclined to cause trouble. Goats, as the writer Brad Kressler (2010) puts it, seem to have a disdain of "human dominion" and an intelligence that allows them to disrupt and challenge people's attempts to control them.

"Myor bakkar bahute haraimi chhan [my goats are real bastards]," my friend Kamla complained to me one day, rivulets of sweat running down her face and body from the effort of chasing after her goats who had strayed into a neighbor's field covered in green stalks of wheat. She had been teaching me how to knit, a favorite pastime of the grazing grounds in Pokhri, when the goats had disappeared, forcing Kamla to run after them. I had been instructed to wait with a stick at the point where the goats would try and make a run for the orchards that lay above the sliver of road that separated them from the grazing ground. When she returned after retrieving the goats, we sat down under a tree and poured sticky-sweet tea from a flask into steel tumblers. As the goat with the glossy black coat passed us, clearly unrepentant, she smacked him with the side of her sickle. "Always causing trouble," Kamla grumbled, "spend the whole day running here and there after them. Look how peacefully the cows are grazing. They'll fill their bellies and come home quietly. No problems with them. But these *dusht jaanwar* [naughty animals] . . . how much they make me run." We watched as the goat she had smacked wandered in the direction of another set of fields. This was a daily routine for him. Unlike the other goats who were happy to browse on thorns, grass, and nettles, this

FIGURE 10. Ensuring that goats don't stray into other people's fields is constant work.

goat was drawn to cultivated fields as if by a siren song. If was a sure bet that he would be found grazing joyously on peas, beans, wheat, and whatever else he could find if left unsupervised for too long.

What remained with me from all the time I spent at grazing grounds was a keen awareness of how much work was involved in raising even those animals who are thought of as low maintenance. I had learned from watching these women how much this labor was performed at the expense of their own health and the welfare of their family. Many women in Kumaon told me that they spent more time caring for their animals than they did their children. Munni Rekhwal, another companion from the grazing grounds, was famed in the village for how much time she spent with her animals. She recalled how even her children would often remark that animals were more important to their mother than they were. She told me, with a mixture of pride and regret, about how little time she had been able to spend with her youngest son in his early years.

> I would tie my three-year-old son to one of the legs of the cot with my *saree* to keep him out of mischief. Then I would go off with the animals during the day. His older brothers and sister would feed him and take him to the bathroom. He would cry and cry until he started choking, but what could I do? The animals couldn't go out on their own . . . my husband was busy with his work. You can't just abandon animals, expect them to roam on their own. I took a vow (*vachan*) . . . raised them like children . . . with more care than my own

children. Cutting oak, cutting grass, chasing around after them the whole day. Once I found myself standing in front of a leopard when I followed one of the goats into the jungle to bring her back home. All this I do for my animals. How can I not feel *mamta* for them?

Munni's words reveal how much the kinship women feel for the animals they raise is rooted in feminized productive practices and care. Relational care, as so many feminist scholars have pointed out, is inextricably intertwined with routinized, arduous labor (Duffy 2007).[18] It is through arduous everyday acts of labor for and on animals that women, and a few older men who are sometimes handed the responsibility of caring for the family's animals, come to experience feelings of love and kinship for the animals they raise. The very fact that the term often used to evoke this sense of kinship is *mamta*, or maternal love, speaks to the gendered division of labor that is at the heart of this love. While this does not at all diminish the affective intensity of this connection and the shifting and contingent ways in which it is made real in the lives of humans and animals, it does illuminate the patriarchal contexts and conditions in which interspecies love comes to fruition.

This labor is not just gendered, it is also fundamentally "intercorporeal" (Alaimo 2010; Al-Mohammad and Peluso 2012). I was viscerally reminded of the embodied nature of care on the day I watched Bimla *chachi* help her goat with giving birth. The doe had already given birth to two kids with ease

FIGURE 11. A few older men also claim kinship for the goats they raise.

but was struggling to push the third out. We could see the head of the third kid through the amniotic membrane, but the rest of the body did not appear despite the fact that the doe had been straining to push it out for twenty minutes. At first *chachi* tried to tug gently on the head, side to side and then up and down, in time with the doe's contractions. When that did not work, she gently coaxed the doe into lying down on a pile of grass. She then worked her fingers gently around the doe's vulva to loosen it up while simultaneously pulling on the head with her other hand. I watched with trepidation, worrying that she would injure the kid or the doe, but her touch was infinitely tender. After a few tense minutes, the kid finally emerged out of the membrane and fell to the ground. The doe's body finally slackened. *Chachi* then placed the kid in front of the doe, who licked her clean. Her hands were covered with slime and blood, a visceral reminder of how she had opened her own body up to the other bodies in this encounter.

We were all affected by the other in that moment; our bodies were permeable and open to potential interconnection. Each of us was moved to participate in the life of another—the doe, who bathed her newborn kids with her tongue, sustaining life; *chachi*, whose breath was labored from the effort of crouching beside the doe in the unventilated shed but had refused to leave her side during the long labor; the second kid, who snuggled close to the first for warmth and companionship; and even I, as I kept guard over the other two kids to make sure they did not get too close to the cows. That moment had engendered a density of embodied affect. We were each, in Sara Ahmed's (2010, 24) words, "oriented towards" the bodies around us and had taken up residence in each other's "bodily horizon." This encounter, which was shot through with pain, pleasure, anticipation, and reliance, had shaped each of us—our trajectories, orientations, and desires—as well as the world that was taking form around us. Thinking about that day in the shed a few years later, I recognized how it was through the "traversal of affects" such as these that different beings came to stand in relation to one another as kin (Singh 2011, 444).[19]

I use the phrase "in relation to one another" to deliberately emphasize that this sense of kinship was not one sided but reciprocated by the animals who were the recipients of this embodied care. When I returned to *chachi*'s house in 2012, a year after the birth, she laughingly told me that the kid she had helped bring into the world now treated her like a second mother. When I accompanied *chachi* to the grazing grounds that were thickly carpeted with grass, I observed how the yearling, whose cream-colored coat was dabbed here and there with brown patches, remained close to her the whole time that we were out. Every time *chachi* moved away from the goats to gather some firewood or move into the shade, the yearling would leave the rest of the herd behind and follow her. The humans sprawled under a tree laughed as she attempted to climb up an oak tree after *chachi*, who was cutting firewood for the evening. Every time the yearling's front hooves slipped off the trunk, she would bleat loudly in frustration. "I'm coming," *chachi* shouted. "Doesn't leave me alone," she grumbled from among the branches, "she's become very naughty." But I could tell that this affective proximity was a source of joy to her. When she finally slipped down the tree with her usual grace, the kid was delighted and skipped around her. She would not stop jumping up to *chachi*'s stomach, the highest she could go, until the latter bent down and scratched her head. Over the next few days, I noticed the goat kept *chachi* firmly within bodily orbit, nurturing the proximate intimacy that had been born in the shed a little more than a year ago. To borrow again from Sara Ahmed (2010, 28), the goat was gathering and incorporating Bimla *chachi* into her "near sphere," keeping within reach the person who gave her pleasure and comfort.

I was struck by a common thread in such instances of intimacy between people and the animals they raised—the fact that humans and animals could "speak" to one another in ways that transcended their species difference. As I discussed in the introduction, animals were able to communicate with each other and with humans in ways that clearly "revealed the presence of a mind and a soul" (Coetzee et al 1999, 101). The yearling was able to signal that she wanted her head scratched by jumping on Bimla *chachi*. Kamla's glossy black goat would respond to direction only if pulled by one of his horns. These were forms of "embodied communication" (Smuts 1999) made mutually intelligible by the fact that the bodies in play remained open and oriented to one another over the course of quotidian acts of care and labor.[20] This communicative connection was the basis for the intersubjective nature of these relationships, relationships characterized equally by risk and reward.

The Event and the Ethical Everyday

Let us return now to the sacrificial arena. The everyday forms of relatedness that emerged between humans and goats play an important part in making the sacrifice of animals ritually and affectively meaningful and powerful. Goats, as I have argued, are precious animals, raised with sweat and tears through gendered labor. Offering them to *devi-dyavta* in sacrifice has a twofold effect—first, it deepens relatedness between those who offer the sacrifice and the being that is sacrificed; and second, it indexes willingness on the part of a *bhakt* to give up something valuable in the service of deities, thereby confirming the intensity of the devotional ties that bind deity and devotee.

Let me start by considering the first form of relatedness—between sacrificial victim and sacrificer—and the forms of loss and ethical practices it engenders. On one of the days of the Navratri in 2010, I was visiting a *devi* temple perched high atop a steep hill when I noticed a young woman with red-rimmed eyes and tearstained cheeks sitting beside one of the smaller temples in the same complex as the main shrine. I sat down next to her and asked if she was all right. A pair of red-billed blue magpies chattered angrily at one another in the tree above us before flying away into the valley, their white ribbonlike tails touched gold by the sun. She explained that her family had sacrificed a goat that morning, one of three goats she had raised herself. "I'll feel terrible tomorrow," she said with a wry smile. "The cowshed will be so empty. But this is *real* [*asli*] sacrifice." When I asked her what she meant by "real sacrifice," she responded with a question of her own. Did I know that the gods once used to demand human sacrifice? I nodded. In those days, she continued, children used to die to save their parents; they went willingly to their death so that everyone else could be safe.

> Today things are different in some ways. Nobody sacrifices children anymore; we sacrifice animals. But in some ways it's not that different. When you live with animals and raise them, they are like your children. And when they shake [to consent], you feel that sense of pride in knowing that your children will die willingly to repay the *mamta* they receive. That's a real sacrifice, when you offer a goat you love almost as much as your child, and he repays your love by dying.

For this young woman, sacrifice closed a circle of kinship that began with the experience of living alongside and laboring on animals. As Anna Tsing (2012) notes, "domination, domestication, and love are deeply entangled."[21] Like other women and some older men, this woman, too, spoke of feeling *mamta* for the goats toward whose lives she had oriented her own. Her feeling of relatedness to the goat she had raised was grounded in acts of labor that required her to align herself in relation to other bodies; an alignment that I have argued was saturated with embodied forms of affect. This already powerful sense of relatedness was intensified by the death of the animal in sacrifice. For the young woman, the goat's sacrificial death cemented her belief that she and the animal were united by a kin relationship. As she understood the situation, the goat had repaid the maternal love with which she had raised him by dying on behalf of his human family. This was also a claim that Neema had made in response to Girish's accusation that she did not really care for the animals whom she had offered as sacrifice on that Navratri day in 2011.

What Neema and the young woman invoked was a bond of "reciprocal indebtedness" similar to that described by Sarah Lamb in Bengali families. Lamb (2000, 46) writes that focusing on these transactional relationships provided one of the main ways for "people to speak about the connections binding the generations." Like the Bengali families that Lamb (2000) worked with, many villagers in Uttarakhand, especially women, understood sacrifice as a transactional relationship wherein a lifetime of care and nourishment creates a *karz* (debt) that the sacrificial animal pays with his blood. But unlike forms of intergenerational nourishment in human family relations that are fairly linear, in the sense that the responsibility of giving passes from parents to children, sacrifice entails a more circular form of indebtedness. This difference is owing to the nature of repayment expected by "children." Human children must "provide for their parents when they become old and ritually . . . [nourish] their parents as ancestors after death" (Lamb 2000, 46), but they do not have to die to return the debt they owe their parents. In the case of sacrifice, however, the "debt" is repaid by animal "children" with blood. The spectacular nature of the offering now imposes a debt on humans in turn. They must acknowledge this final and unmatchable act of kinship with an offering of their own.

What people offer in turn are ordinary, uninnocent ethical acts that derive meaning and consequence from the fact that they emerge through remorseful, sometimes even haunted, reflection on past and future instances of violence. These ethical practices are fundamentally relational; work on the self is already and always work on the self in relation to another (Al-Mohammad and Peluso 2012). Following Veena Das (2007, 134; cf. Lambek 2010; Laidlaw 2014), I think of the ethical as a "dimension of everyday life" wherein moral selves are cultivated through an engagement with the stuff of daily life. "Such a descent into the ordinary," Das clarifies (2007, 134), "does not mean that no attempt is made to work on this ordinary in the sense of cultivating critical attitudes toward one's culture as it stands, and also working to improve one's condition of life, but that such work is not done by orienting oneself to transcendental, objectively agreed-upon values but rather through the cultivation of sensibilities *within* the everyday."

Indeed, people's practices of mourning and atonement in Kumaon did not emerge from any societal consensus on the (im)morality of sacrificial violence. Most often, they were fleeting and unremarkable—a quick caress of a goat's head before he was taken to the kill area; the flinch of a body when the sacrificial blade came down to sever head from neck; tears streaming down a face at the end. But through such ethical gestures, people reaffirmed their connection to the animals whose deaths they mourned and extended relatedness into realms beyond this life. In other words, ethics and kinship were constitutive of one another. It was only kin of a kind who could be mourned with varying degrees of remorse, but the act of mourning was crucial in deepening and extending that kinship. As Thom Van Dooren reminds us, "mourning is about dwelling with a loss and so coming to appreciate what it means, how the world has changed, and how we must ourselves change and renew our relationships if we are to move forward from there."[22] By dwelling, even if briefly, in loss, people reflected on how to come to terms with their responsibility for loss. Such reflections on responsibility were the starting point of cultivating what Donna Haraway (2016, 2), drawing on Karen Barad, calls response-ability, noninnocent "reflections on who lives and who dies and how."

The compassionate gestures, moments of love, and acts of care that constituted these ethical response-abilities were woven into the fabric of everyday social life. A young woman who was feeding a young kid shelled peas responded to my question about what he had done to deserve this rare treat with the comment that her family had dedicated him to a *dyavta* and that she wanted him to live a good life before he was sacrificed. An old man kept a notebook in which he made a little mark to represent each goat his family had sacrificed so that he could account for exactly how many deaths he was responsible. A *guru* told me that he muttered a little prayer for the soul of a goat before sacrificing him during midnight *pujas* to Masan, the lord of the cremation ground who comes in many different forms.[23] A woman in her twenties told me she still wept every time she thought of the different goats that she had raised and then seen sacrificed; she said that she had taken a vow to never sacrifice an animal again. These were all small but significant acts of care that, without "calling undue attention to themselves" were directed toward other bodies (Lambek 2010, 2). At the heart of all such gestures was an acknowledgment that one's life was intertwined with that of another and that this connection implied some responsibility on the part of those who were alive to those who were or would soon be dead, especially because that death was intended to benefit the living.

One such act is emblazoned in my memory for how it speaks to the endurance of these interspecies entanglements. In December 2014, I accompanied Gita, a twenty-three-year-old college student who had become a close friend, to a temple in Pokhri dedicated to Airy *dyavta*, the protector deity of livestock, particularly cows. When I pressed Gita as to why we were going to the temple when the cows had not given birth and were not sick, she put a finger to her lips. "I can't tell you yet," she said. "But I'll tell you after we go to the temple." When we reached the temple, a humble stone shrine with a spectacular view of the Himalayas, which were revealed in their full glory that day, Gita folded her palms and prayed silently for a few minutes. She had brought a little bottle of oil, a wick of cotton, and an earthen *diya* with her. I watched as she carefully

FIGURE 12. The airy *mandir* in Pokhri.

dipped the cotton in oil and then lit it with a match from the box she carried in the pocket she had sown into the inside of her *salwar*. We were sitting under an oak tree festooned with red banners and bronze bells, offerings to Airy *dyavta*, who was usually a meat-eating deity but, in this village, did not take sacrifice. On the plot of land just below the temple, a family from the plains had built a lavish second home. I could hear their Tibetan sheepdog barking, his deep-throated voice echoing through the valley. Gita put the lit *diya* in one of the three arched niches. Each contained five little stones marked with *pithiya* and grains of rice. These stones represented Airy. Next to the *diya*, she put a few *batashas* (sugar drops) as *prasad*.

I was increasingly curious as to the meaning of this private ritual. As we made our way back down the narrow rocky path to the road, I asked Gita who and what she had prayed for. "I didn't want to tell you before, because then it wouldn't have come true," she said.

> I was praying for a goat that our family had sacrificed many years ago. I was young then, nine or ten years. I was very sad when he died. I cried for many days . . . every time I went to the shed. Then last week this same goat came to me in a dream. I opened the door to the shed to take the animals out to graze. And he was sitting there next to Meena [their cow]. I felt very anxious. I was his favorite [among the mother and sisters] in the family. He must have been trying to tell me something. I talked to my mother and she suggested I do a small *puja* at the Airy *mandir* (temple). That's why I came. I prayed that his soul would find peace . . . and that he be born well in his next life.

Gita was more than a little embarrassed as she revealed what had brought her to the Airy temple that day. I assured her that I did not think her impulsive *puja* was silly. I asked if she thought of the goat often. "Not often, no," she responded. "But if I go back to that part of the forest where we used to sit together, I think of him. When I walked to the Shivratri mela in Kapileshwar last year, I passed Bhalu ki Gadheri. The ground there was covered in *buransh* (rhododendron) flowers. I thought of him. He loved eating them. It was difficult to bring him home in that month. That's how I remember him, now and then."

As Gita spoke, I was impressed by how the brief entanglement of her life with the goat's had created a connection that lingered, albeit in tenuous fragments, so many years into the future. Gita's *puja* was an ethical act that was prompted by this connection and its violent severance. Her dream, in which she briefly reencountered the goat in the space of the shed, was for her a real "interlocutory possibility," opening up a "dialogical in-between space, a space in which the living and the dead can meet" (Mittermaier 2011). The dream called on her to open herself up once again to the now dead goat. It was also a forceful reminder of the fact that she was not an isolated, autonomous self but *existed in relation to* another. Her *puja*, an ordinary, mundane practice for young women in this region, was a small gesture full of concern and hope; concern that the goat's soul was in unrest and hope that he would find peace and be reborn. It was also an act that was limited in its scope. Gita, although she felt sorrow at the goat's death, believed that it had been necessary for him to be sacrificed. However, within this larger world of violence and destruction, there was more than a little room for everyday gestures that allowed people to dwell in and on loss, reciprocity, gratitude, and love.

Of late, several scholars have argued that the emphasis on the ordinary and everyday in the anthropology of ethics runs the risk of diluting the specificity of the ethical by locating it throughout the fabric of everyday social life. When ethics are this ubiquitous, the critics ask, are they even ethics anymore? It is this concern about the attenuation of the ethical as a distinct sphere of life that prompts Jarrett Zigon's (2014, 750) claim that ordinary ethicists are unable to provide an approach for recognizing what counts as ethics because for them "everything in social life is ethical." Zigon argues that ethics are not intrinsic to language and social activity, but emerge from specific "ontological conditions for being in the world" (Zigon 2014, 762). While I am sympathetic to Zigon's claim that our definition of ethics should rely on the particular conditions of their emergence, I believe that these two ethical registers—the everyday and the "eventful"—are mutually constitutive rather than opposed (Lempert 2014).[24]

Even though the ethical acts I have just described were indeed ordinary and sometimes even tacit, they emerged in response to spectacular acts of violence that were out of the ordinary. Sacrificial violence spurred moments of critical self-reflection for those who judged their own or others' actions as even faintly immoral. The uncertainties raised by such explicit ethical co-nundrums resonated through the future, igniting a series of ordinary ethical acts that hearkened back to moments of breakdown. Gita's intimate *puja* was sparked by a dream that brought her back to a moment of moral and per-sonal anguish. The eventful was thus temporarily resolved in the everyday. Obversely, it was the intensity of everyday doings of ethical interspecies kinship—enacted through practices of care and love—that made the sacrifi-cial moment such a compelling occasion for self-reflection. When people like Gita mourned the death of sacrificial animals, they did so because everyday acts of regard and attunement had created deep bonds between them. Gita grieved the goat not just because he had died a violent death but because she had shared a strong prior connection with him. "I used to take the animals out to graze earlier than usual in the month of Phagun," Gita had said. "That way he had more time to eat the buransh." Gestures as ordinary as these were what gave rise to extraordinary ethical reflection. The everyday and the eventful do not stand in any linear relationship to one another but constitute one another.

In thinking about the ethical as emerging through the interconnections of the eventful and the everyday, I am reminded here of Veena Das's (2013) obser-vation that in Vedic texts, sacrifice forces a melancholy awareness of how "life is conjoined to death." These texts, she argues, "make us think of our forms of life as entailing the inevitability of death and the inevitable contamination of our everyday life by the violence we commit." It is this gloomy awareness that forces the texts to offer some "respite" through instructions on inculcat-ing "'noncruelty' in inhabiting the world with animals." In Kumaon too, the awareness that the future holds violence encourages people to generate forms of care and reciprocity in the present, a claim I return to in the next chapter. It is in this complex relatedness of human and animal lives, which precedes the sacrificial act but is also constituted anew through it, that such small but significant ethical acts flourish.

Sacrifice as Savagery

Some people in Uttarakhand are not convinced that such gestures constitute an ethical practice. They point to the violence of the sacrificial act and argue that no ethical orientation is possible in the face of such intentional destruc-tion. For these opponents, it is not just ethical responsibility as a human that is at stake but also the question of what it means to live ethically as a Hindu (Govindrajan 2014). Indeed, as I mentioned in the introduction, debates in the public and legal sphere, even when they have centrally engaged with the question of what it means to be ethical toward animals, have taken place within a larger framework of whether sacrifice is actually sanctioned by Hinduism or constitutes a deviation from the "authentic" religion. Their definition of Hinduism is based on a narrow interpretation of Vedic texts and excludes locally meaningful forms of devotion and practice.[25] Thus, these narratives of reform discount not just villagers' everyday ethics in relation to sacrificial animals but also their understandings of "reciprocal indebtedness" vis-à-vis their *devi-dyavtas*; the idea that sacrifice is the due of local deities in return for their benevolence and favor finds little support among opponents of the practice.

The matter of defining legitimate Hindu practice came to the fore when demands for a legal ban on animal sacrifice gathered force from 2010 onward. In the Public Interest Litigation (PIL) filed by the PFA, the petitioner sought the High Court's intervention on behalf of the "poor dumb animals whose lives are mercilessly sacrificed in the name of religion and faith." In response to section 28 of the Prevention of Cruelty to Animals Act, which states that nothing "contained in the Act shall render it an offence to kill any animal in a manner required by the religion of any community," the petitioner declared that sacrifice "is . . . 'not required' or 'unnecessary' in Hindu religion . . . [and] is also prohibited according to the Hindu religious texts. There is absolutely no mandate in Hindu religion to kill animals. . . . Regard for all life forms and 'ahimsa' is the foundation of the Hindu religion."[26]

Thus, even when articulated by animal-rights groups who were dedicated to raising the urgent and unavoidable question of what it means to live ethically in relation to sentient beings, the case against animal sacrifice still relied on the separation of good religion from bad religion, true belief from false belief. When I talked to her over the phone a few weeks before she filed the PIL, Gauri Maulekhi of the PFA rejected the argument that sacrifice was an important part of local ritual practice and argued that "tradition" was merely a fig leaf for societal violence. Sacrifice was a mockery of Hindu spirituality, she continued, and had to be eradicated in much the same way as other unpalatable superstitions such as sati or child marriage.[27]

By pointing to how animal-rights discourse and practice is sometimes framed in terms of a distinction between good and bad Hinduism, I do not wish to discredit the compelling affective, ethical, and normative claims made by animal-rights activists. In a powerful article on animal-rights activism in India, Naisargi Dave (2014) describes how activists work tirelessly to address the issue of violence against animals. Their daily acts of witnessing unbearable violence against animals, she argues, "constitute an animal event in tethering human to nonhuman, expanding ordinary understandings of the self and its possible social relations, potentially blowing the conceit of humanity apart." In Uttarakhand, too, activists must constantly participate in such potentially transformative "animal events" as they go about their daily work of witnessing and battling violence against animals.

And yet despite activists' claims that they are concerned only with justice for animals, their liberal rights discourses are also often troubling commentaries on the relationship between religion, modernity, and the nation. In 2016, for example, Gauri Maulekhi participated in a televised debate about whether Eid could be celebrated in an "eco-friendly" manner, that is, without the sacrifice of animals.[28] In the debate, she argued that all Indian religions, except for Islam, had shed the burden of superstitious practices and exhibited signs of progress over time. As an example of Hinduism's progress, she talked about the High Court's order restricting sacrifice in Uttarakhand. Maulekhi argued that the success of the campaign against animal sacrifice in Uttarakhand revealed that Hindus were essentially a progressive community, willing to discard pernicious superstition for the sake of the nation's progress. There was only one religion, she declared, that consistently flouted the nation's laws with impunity, and that was Islam.

Maulekhi's excoriation of Islam as a religion that sanctions unimaginable cruelty against animals is echoed across the country by right-wing Hindu nationalists and religious reformers. Their attempts to reform and reconstitute Hinduism by returning it to its uncorrupted past cannot be seen in isolation from the desire to establish Islam as its ignorant and superstitious Other. Hinduism must be modern because Islam is not. Appeals by petitioners who seek to constitute a modern and reformed Hinduism by eradicating what they argue are superstitions (derived from Islam and Christianity) that corrupted Hinduism have been crucial in persuading the judiciary to act against animal sacrifice across the country. For instance, in response to a 2013 PIL filed by an animal-rights activist, the High Court of neighboring Himachal Pradesh ruled that such rituals "must change in the modern era." The abandonment of superstitious practices such as sacrifice and the embrace of an authentic Hinduism have thus become conditions for full citizenship in a modern nation. There is no place in the modern and secular temporality of the nation-state for practices that are thought to belong in the past.

In 2011, the High Court of Uttarakhand responded to the PFA's petition and announced that animals could now be killed only to provision food and not with the primary purpose of pleasing deities. The court laid down two stipulations to make the scope of its judgment clearer: first, as I mentioned earlier, sacrifice would no longer be allowed on temple premises and could only take place in slaughterhouses where available, and if not, in sheds or areas set aside for the purpose elsewhere; and second, all meat from an animal that had been slaughtered had to be consumed by those who had killed it. The second stipulation, to all intents and purposes, put an end to buffalo sacrifice in the region, a practice that was already severely curtailed by public health laws by the time I began fieldwork. Upper castes believe that the consumption of buffalo meat is polluting and associated it with Dalits, particularly with *jatis* whom they regard as "untouchable."[29] Even those low castes who consumed buffalo meat in the past refused to do so at the time of my fieldwork; indeed, they claimed that only Muslims eat buffalos. I was told by villagers and temple officials alike that the refusal by members of any caste to consume this meat over the last few decades had resulted in the carcasses of sacrificial animals being thrown into a ravine, where they would rot and spread disease. The High Court's order that the meat of a slaughtered animal would have to be consumed by those who had killed it made such forms of disposal difficult to sustain.

During my time in Kumaon, I witnessed very few buffalo sacrifices. Temple administrations were afraid that buffalo sacrifice, which is a decidedly grander spectacle than the sacrifice of goats, would attract the ire of animal-rights activists, religious reformers, and state officials; as a result, long before the court's order, many temples in Kumaon, such as the Naina Devi temple in Nainital, had begun turning away devotees who brought buffalos for sacrifice (Govindrajan 2014). Priests and temple officials, even those who acknowledged that goats had been sacrificed on temple land for some time even after the court's ruling, told me that buffalo sacrifice had stopped almost entirely, except perhaps at a few temples in faraway districts such as Pithoragarh or in smaller village temples. "It's too difficult to hide," a priest at a major Golu temple in the region confided, "and it's not worth the risk of allowing it on our premises. The animal-rights activists will file a case against us, and then we will have to go to the courts."[30] As a result of this complicated social, political, and legal history, by the time I started fieldwork it was rare to see buffalos sacrificed in major temples on festival days.

The court's order made it difficult for devotees to sacrifice even goats. A year or so after the order was passed, most temples had started ordering devotees to kill sacrificial animals themselves in a place adjoining but not inside the temple. The temple dedicated to Golu *dyavta* at Chitai, which had been a prominent target of the PFA campaign against sacrifice, announced that devotees would now have to bring their own priests, *puja ki thali* (plate of offerings), and *banyats* (cleavers) if they wanted to sacrifice a goat. Moreover, the animals would not be allowed to enter the temple premises but would have to be sacrificed in a field outside the boundaries of the temple. The situation changed quite dramatically in 2016, when, in response to a PIL filed by a resident of Nainital, the High Court of Uttarakhand permitted devotees in Nainital to sacrifice animals during the Nanda Devi festival in September; the animals were dedicated in the temple and then sacrificed in a separate area. However, despite this turnaround, the issue of sacrifice remains contentious, and it is unclear what turn legal debates will take in the future.

For many *paharis*, the critical question raised by the High Court's decision was how *devi-dyavtas* would respond to what was clearly an infringement of their sovereignty. The court's order had intervened directly in the workings of a world where bonds of mutual reciprocity between humans, animals, and *dyavtas* were crucial to maintaining the ritual and material prosperity of the *devbhumi*. While some *paharis*, including those who had migrated to cities, supported the court's decision and believed that sacrifice was a backward and cruel practice, others were concerned about angering *dyavtas* by denying them their due. Several temple administrations responded by filing counterpetitions of their own demanding that people be allowed to sacrifice animals to deities in keeping with what they described as age-old tradition that was crucial to *pahari* social reproduction.[31] The idea that the court's curtailment of sacrifice had torn apart a moral economy of reciprocity was articulated forcefully by Puran Bisht, one of the founding members of the Gwal Sena, when I interviewed him in 2011. In the course of a larger conversation about what he perceived as a concerted attack on *pahari* culture, he was quick to point out that the court's decision to disrupt the relationship between *paharis* and their *dyavtas* would have devastating effects. "These practices were started by our ancestors, they are written about in the Dharmasastras," he told me.

> Ever since they [the PFA] stopped the sacrifice at Devidhura [a major temple against the administration of which the PFA had filed a police report] last year, there has been almost no rain in that region. There is nothing to eat in those villages. People are starving. [This is a] natural disaster [*prakratik aapada*] and more will follow. There will be human sacrifice in place of animal sacrifice now. Nature and the gods of this *devbhumi* demand blood. If they don't get it, they show their angry form. Thousands of people will be finished [by them].

The belief that the eradication of sacrifice would result in a return to the dark days of *narbali* was pervasive and widely accepted even by those villagers who felt uncomfortable with their role in sacrificing animals (Govindrajan 2015a). On a rainy afternoon in July 2015, I was sitting with Prema Pant and her three daughters, who were all waiting for the downpour to stop so that they could go about their daily chores of washing the dishes and cutting grass for the cows. Our idle chatter, fueled by endless cups of sweet tea, turned to the neighbors who had offered a sacrifice as thanks for their newborn grandson a few months ago at a temple dedicated to their *kul devi* in Almora district. Prema told me that the temple had recently created a separate space for devotees to sacrifice animals. When I asked what had motivated the temple administration to do so, she told me that the decision had been made after a leopard terrorized the town of Almora one day. Among the people attacked by the leopard in the middle of the day in the busy space of the bazaar, she told me, was the son of one of the head priests at the temple. The attack, she declared, was an unmistakable sign of *dyavta prakop* (divine rage). The leopard, Nari Ram, another villager, told me a few weeks later, had a "god gift" and was tasked with bringing wayward human devotees back to the fold. While some were skeptical of these rumors, to many others it was clear that this series of seemingly disconnected events were a powerful demonstration of the sovereignty and power of local *dyavtas* who were displeased by the High Court's attempts to disempower them by interfering in their relationship with their devotees. Thus, "modernity," insofar as it involved the abandonment and outlawing of "traditional" practices, would not come without a significant price. Similar beliefs about the sacrifices extracted by modernity emerge in different form in other parts of India. In Bhilai, the site of one of the largest steel plants in Asia, Jonathan Parry (2008) found that many workers firmly believed that the high volume of industrial accidents were a result of "Kali's wrath [*devi ka prakop*] and her way of asserting her claims" to the legitimate blood sacrifice of animals that she had been denied by her devotees.

Given the power of this moral economy where the intercession of deities on behalf of their human subjects exists in a delicate balance with devotees' abilities to perform acts of propitiation for their *devi-dyavtas*, it is unsurprising that ritual sacrifice remains an important thread in the fabric of life in the mountains despite ongoing legal and political attempts to curtail it. In 2015, I watched from the shade of a pine tree as a steady stream of young men flocked to an open field just outside the walls of Chitai temple in Almora and sacrificed the goats they had brought with them. The men used their own cleavers, and many were accompanied by priests from their village who chanted the necessary invocations before the goat was summarily beheaded. A number of women watched from a safe distance. When I asked the priests at the temple if they still blessed the animals before the sacrifice, I was told that the temple no longer participated in sacrifice in any way. One of them, however, asserted that "Golu's power is everywhere. Whether they sacrifice the animal inside or outside, Golu will still accept it. This whole area belongs to his court [darbar], not to the High Court."

His words were echoed by Jeevan, one of the friends I was traveling with, who explained that one had to adhere both to the sovereignty of the High Court as well as the sovereignty of devi-dyavtas.[32] Jeevan told me that while the state had its own place, the gods could not be ignored either.

> We can't offer puja in temples anymore, but this whole area [ilaaka] belongs to our devi-dyavta. So we take their name, and cut the goat. Doesn't matter where you do it, this whole area belongs to them. That's what these animal-rightswallahs do not understand. We have a duty to our dyavtas.

On Love and Violence

The recent movement against animal sacrifice has raised difficult but crucial ethical questions. It is no longer possible for people who sacrifice animals to do so without an urgent and painful awareness of the violence that they are inflicting on these creatures. Some are unable to reconcile their ethical aspirations of the self with this violence. In 2014, Dewan, a young man from the Tamta jati whom I had known for a few years, declared that he would no longer participate in any rituals that necessitated animal sacrifice. He was troubled by the potential consequences of this decision, but was firm. "How long," he asked me, "can we justify chopping the heads off innocent animals in the name of our dyavtas?"

> Maybe they did demand animal sacrifice once. Maybe terrible things will happen if we stop sacrificing animals. But we have to think beyond that. We say that these animals consent to their sacrifice. But do they really? I don't know. Even a dog will shake if you sprinkle water on it. . . . Religion must change. People did things in the past . . . in the name of religion . . . they did wrong things. The way we Harijans were treated, are treated, was justified in the name of Hindu religion. . . . I care about one thing alone: if I am snatching someone else's life for the sake of my happiness, can I truly be happy?

When I left Dewan, I was troubled by what he had said. Like him, I had found it hard to come to terms with the violence that sacrifice entailed. The first time I went to a festival where animals were sacrificed on a large scale, I was stunned by the sight of hundreds of severed goat heads with yellow eyes staring up at me. When I first lifted the camera to my eye, the sea of blood in my viewfinder made my hands tremble. Given my own visceral reaction, I was curious to hear how people felt about their role in this spectacular display of violence. I wanted to take seriously people's religiosity, but I was also concerned about falling into the trap of essentialist and Orientalist understandings of religious identity. I was determined to resist narratives about the innate cruelty of rural and lower-class people who are thought to be incapable of understanding that animals are worthy of care, love, and, indeed, life. Perhaps unintentionally, activists often imply that animal rights is the natural domain of a cosmopolitan elite who must lead others into the light. This was a claim that I wanted to distance myself from even as I sought to understand why people sacrificed animals.

Over time, as I immersed myself in the social life of these villages, I learned that the meaning of sacrifice was too complex to be captured by any of the narratives authorized by reformers. From those who told me stories about *devi-dyavtas*—their histories, their desires, but also their effects on the lives of individuals and communities—I absorbed the knowledge that people existed in a delicate balance of mutuality with *dyavtas*. This was a relationship characterized by a "reciprocal indebtedness" wherein devotees had to try and repay the gods for all that they had been given by virtue of the latter's grace and generosity. Reciprocity with *dyavtas* was not thought to be restricted to people; animals, too, were part of a social world of care and mutuality. People told me about leopards—devotees of *devis*—who would fast during the Navratris. They claimed that leopards did not hunt during that period. This was a world sustained by delicate checks and balances, and it was a world that reformers glossed as superstitious.

After spending hours with people and animals at grazing grounds and in homes and sheds, I learned that raising animals requires multiple bodies to orient themselves in relation to one another. As different beings opened up to one another—a risky and rewarding process—their lives came to be entangled in ways that were rich with the possibility of care, concern, mutuality, and even love. It was through these entanglements that people and animals were constituted as kin. I was taught that this kinship had its own language, one that I came to speak haltingly and joyfully even if I was not always understood by people and animals alike. It was this kinship, I was told by so many, that gave meaning to ritual sacrifice. It was pain at the loss of a being who was dear, whose presence in one's bodily horizon was a source of pleasure, that made sacrifice *truly* a sacrifice.

From witnessing and hearing about the small gestures of care, remorse, absolution, and love that people made in the wake of an animal's death, I came to see that violence could generate ethical ways of engaging with the world. I recognized that these ethical practices were modest in their scope, but I was still moved by their sincerity and their power. These were intimate ethical practices rooted in the everyday that derived their meaning from people's understanding of their lives as inextricably entangled with those of others. I saw that sacrifice generated ethical orientations and reflections in and on relationships and subjects—both living and dead—that lingered long beyond the fleeting sacrificial act.

One day I asked Bimla *chachi* whether she thought the practice of sacrifice would stop entirely in the years to come. She was circumspect about the future but unwilling to concede that sacrifice was as bad as "those people from the Gayatri Pariwar say it is." When I told her there were people who loved animals and believed that sacrifice was a cruel practice sustained by those who didn't really care about animals, she was visibly disturbed. "Why do they think that we do not know about love?" she asked. "Have those people ever brought *pathiyas* (kids) into their home because they were worried that the leopard who came every night would eat the goats? Have they ever pounded *haldi* (turmeric) and applied it to a festering wound every day for a month?" She looked down at the kid who was nuzzling her side, the same kid that she had helped bring into the world, and then looked back up at me. "Is this not love?"

The Cow Herself Has Changed
Hindu Nationalism, Cow Protection, and Bovine Materiality

Approaching the Sacred Cow

In 2007, the BJP, which controlled the state government in Uttarakhand at the time, enacted a law banning cow slaughter in the state. The Uttarakhand Protection of Cow Progeny Act (Uttarakhand Gau Sanraksha Adhiniyam) forbade the slaughter of cows, bulls, and bullocks of any age. The sale and possession of beef products was declared illegal. If caught breaking this law, offenders would be fined and faced a period of imprisonment of anywhere between three and ten years. The transportation of cattle for the purposes of slaughter was declared punishable under the law. Those who offered cattle for transport and those who were involved in the transportation of cattle would also be fined and were liable for up to three years in prison. Recognizing that a ban on cow slaughter would create a large population of stray cows who were uneconomical to support after a certain age, the law instructed state government and NGOs to establish *gaushalas* (cattle shelters) for the care of "uneconomic cow progeny," a category that included "stray, infirm, disabled, diseased or barren" cows.

In passing this law, the BJP government fulfilled one of the key promises it had made to voters in its preelection manifesto. The BJP's legislative attention to cow protection is in keeping with larger Hindu nationalist discourse and practice whereby cows are figuratively represented as *gau-mata*, the cow mother of a Hindu nation.[1] The mobilization of this "dense metaphor" has traditionally been an important part of Hindu nationalist attempts to constitute an Indian nation that is Hindu in spirit and essence (Pinney 2004).[2]

However, a few years after the promulgation of the law, it became apparent that the semiotically thick symbol of the cow was unable to lend force to

Hindu nationalist projects of cow protection at least as far as putting an end to the practice of cattle smuggling and slaughter in the region went.[3] During the years of my fieldwork, the ban on the transportation and slaughter of cattle pushed the trade in cattle underground and certainly curtailed it, but did not succeed in entirely eradicating it. A steady stream of cattle smuggled from the mountains continued to make their way to urban slaughterhouses. Animals-rights groups, local members of Hindu nationalist organizations, and *sadhus* (Hindu ascetics who are usually part of spiritual sects) were dismayed by the involvement of mountain villagers in the smuggling of cattle. The PFA, which declared that putting an end to cattle smuggling and slaughter was one of its main priorities, demanded better policing of villagers and smugglers alike. In 2013, the PFA alleged that ten thousand cows were smuggled out of Uttarakhand every month, joining the millions of cows being smuggled across India's borders to Nepal, Bangladesh, and other countries every year.[4] "Villages," one of its members told a local newspaper, "were doing brisk business" with the active collusion of the police and politicians.[5] In response to the lack of initiative displayed by "corrupt" politicians, the PFA began to conduct its own raids on illegal cattle markets, sometimes with the support of self-styled *gaurakshaks* (vigilante cow protectionists); the animals were confiscated, and police cases were filed against smugglers.[6] In addition, members of the PFA, acting on tip offs, would often patrol the highways leading out of the state to prevent smugglers from transporting cattle across state lines.

Leaders of Hindu nationalist groups also recognized that the persistence of cattle smuggling and slaughter signaled the complicity of Hindu villagers in the trade. In 2010, Pravin Togadia, the head of the Vishwa Hindu Parishad (VHP), a group that has been at the forefront of violent "ethno-religious political activism" (Jaffrelot 2007) in India since it was founded in 1964, visited Uttarakhand to promote a range of beauty and health products made from cow urine and dung.[7] In a public speech, he drew a connection between rural poverty and cow slaughter, noting that cows were killed because poor farmers did not have the resources to care for unproductive cattle. The solution he offered was the production and sale of products made from bovine waste, an idea that finds a great deal of popular support even among those who do not necessarily align with the Hindu right but are more broadly concerned with issues of rural development. Once people saw how much money they stood to make from selling urine and dung, Togadia said, even cows that were too old to produce milk would become profitable.

Two aspects of Togadia's speech, in particular, are worth reflecting upon. First, his acknowledgment that Hindus across castes sustained the economy of cow slaughter by supplying the cows who were killed complicates the usual

FIGURE 13. A poster in which the head of a local cow protection group exhorts residents of the mountains to support the pillar of the Hindu faith—the cow—and lend strength to the cow revolution.

insistence, in Hindutva discourse and practice, that cow slaughter is the anti-Hindu and, by extension, anti-Indian activity of Muslims, Christians, and Dalits. A number of VHP activists I spoke with also pointed out that the mountain villagers who sold their cows to smugglers were, for the most part, upper-caste Hindus, a constituency they strongly felt should stand united around the cow.[8] Manish, a young VHP *karyakarta* (worker) who ran a store in the city of Haldwani, told me he was disgusted by the behavior of "people who call themselves Thakurs and Pundits but whose actions are worse than Muslims." He despaired for the state of the nation, he said, if upper-caste Hindus would not stand up for their coreligionists and their own mother, the cow. It was clear to him that legislative and social action around the cow would have to tackle the evident unwillingness of upper-caste Hindu farmers to unite around the cow. He and young men like him were not as forgiving of the farmers as Togadia was, and they often spoke of their desire to teach Hindu villagers who supported the trade in cattle slaughter a violent lesson.

The second noteworthy theme in Togadia's speech was his concession that cows who were too old to give milk would somehow have to be rendered economically viable if a ban on cow slaughter was to succeed. Bans on cow slaughter are a crippling blow to the livelihoods (and nutrition) of Muslim and Dalit individuals who engage in slaughter and skinning.[9] While the exclusionary and vicious cow politics of the Hindu right is directed at and has the most dire consequences for these groups, villagers across caste and religious lines who participate in rural cattle-rearing economies have also been affected by the curtailment of trade in bovines. It was the deleterious effect that bans on slaughter were having on rural economies that forced Togadia to acknowledge and address the issue of the unproductive cow. As Christophe Jaffrelot (2008, 12) notes, the "constant oscillation [in Hindu nationalist discourse] between the argument of sacredness and others pertaining to the economic repertoire is very revealing of the fact that, even for the holy cow, the religious motivations were not enough." This is a point of some importance because it complicates the common claim, in both academic and popular discourse, that all upper-caste Hindus are, and have always been, supportive of cow-protection movements as a logical extension of their wholehearted embrace of the religious symbolism of the cow mother

One reason why the 2007 law did not put an effective end to cattle slaughter in the years that followed, then, is that it was at odds with the logics of cattle-rearing economies in which the beef industry is not only an outcome of dairy farming but also supports it by providing a solution to the problem of what to do with unproductive dairy cattle. This is an argument I will return to later in the chapter. However, while this economic reasoning partly explains the lack of upper-caste support for Hindu nationalist projects of cow protection in the mountains, it does not adequately capture the complexity of people's relationships with the *actual* cows they send to slaughter. How do people's embodied relationships with the bovine beings of flesh, blood, and emotion that they help load into trucks going straight to slaughterhouses shape their response to projects of cow protection that depend on banning cattle slaughter? What might a focus on embodied, *real* cows tell us about the reluctance of villagers across castes to support cow protection on the ground even as they revere the symbol of the cow? To put it another way, what might people's everyday kinds of relatedness to different cows tell us about their reluctance to embrace the kinship with *gau-mata* that the Hindu right promotes as morally and materially beneficial for Hindus?

I argue that Hindu nationalist projects of cow protection in Uttarakhand were troubled by their location in "incommensurate worlds" where the undifferentiated and abstract metaphor of the cow mother of the Hindu nation did not sit easily alongside the distinct and lively materiality of the *actual* cows it sought to represent.[10] For mountain villagers, the sacredness of the cow as a mother figure was not a simple matter of abstract belief but was made real in everyday social contexts through their relationships with actual embodied cows. Like many other Hindus, they believed that a cow's body and what she produces, whether milk, dung, or urine, is ritually powerful. However, while recognizing that some cows possessed an undeniable quality of ritual purity and strength, they were also convinced that not *all* cows could claim these qualities. For them, the behavior of some cows, especially the Jerseys who came to dominate this landscape as the result of an aggressive dairy develop-ment program supported by the state, signaled that they did not possess any *shakti* or divine power.[11] This lack of *shakti* meant that people related to Jersey cows in very different ways than to the *pahari* cows they recognized as em-bodying ritual power. It would be easy to read these declarations about the differential *shakti* of different cows and the absence of spiritual merit in raising Jersey cows as a gloss for what is, in essence, an economic choice. Perhaps villagers who sell cows to butchers or abandon bulls by the highway want to conceal, even to themselves, that economic motivations trump spiritual and ritual devotion.

However, such an interpretation overestimates the extent to which social life is human authored. It disregards the possibility that the difference between Jersey and *pahari* cows is not just a human construction but emerges forcefully through the process of human immersion in the life of multiple bovine beings whose dynamic and distinctive materiality *commands* recognition of their dif-ference from one another. The distinctions that villagers described between Jersey *goru* (cow) and *katu* (*pahari*) *goru* were grounded in an embodied, intimate knowledge of the nature of bovine bodies and behavior that emerged through the everyday labors involved in caring for livestock. Through this quotidian immersion in their cows' lives, people came to recognize these ani-mals as differentially material. Bovine materiality thus came to play an impor-tant part in shaping the nature and outcome of Hindu nationalist projects of cow protection in contemporary India.

Over the rest of this chapter, I first trace how the cow emerges as a meta-phorical figure of the Hindu community in Hindu nationalist discourse over the nineteenth and twentieth centuries and then explore her role as an unpre-dictable and creative material force in the lives of those who are called on to form community under the sign of the *gau-mata*. I argue that the metaphor of the cow mother is unable to contain the excess, vitality, and difference of bovine matter, thus making these two (or multiple) cows incommensurable. This incommensurability, I suggest at the end, has unexpectedly powerful effects in the lives of those who have traditionally been subject to violence in the name of *gau-mata*.

Life and Death in the Shadow of *Gau-mata*

Hindu nationalists have long used the cow as an emotive symbol of Hindu community. The "iconic mythopraxis" (Pinney 2004, 107) of the cow as mother of a Hindu nation and embodiment of the Hindu divine is a powerful force in contemporary India and has been mobilized by the Hindu right to constitute Hindu identity and community. The supposed sacredness of the cow as the mother of Hindus, especially upper-caste Hindus, is invoked by Hindu nationalists to mark religious and cultural difference from a whole host of Others. Dalits, Muslims, and Christians, who are even vaguely suspected of cow slaughter, are at the receiving end of heinous, often fatal, violence by Hindu vigilantes. Even as they enact this violence, Hindu right-wing organizations claim that the lack of national legislation protecting this sacred symbol of Hindus constitutes an outrage against Hindus in their own country, a "historical injury" (Chigateri 2011, 140).[12] The fact that India is one of the largest suppliers of beef in the world makes the absence of a national ban on slaughter even more of a scandal for these organizations. While much of this beef is actually buffalo meat, a significant proportion is made up of cow meat.[13] Only a total ban, these organizations argue, can stem this trade that they view as an attack on the religious sensibilities of Hindus.

It is not just Hindu nationalists who represent Muslims as a threat to the cow and the nation. As we saw in the last chapter, animal-rights activists often identify Muslims as animal killers who are ever ready to flout the law of the land. In October 2012, the PFA teamed up with the Uttarakhand Gau Sewa Aayog (Uttarakhand Cow Service Commission), a state body, to begin an ambitious tagging exercise whereby ten thousand cows were tagged and recorded in Muslim-dominated parts of the state before Eid. The exercise was justified on the grounds that the illegal slaughter of cattle spiked sharply just before Eid, when Muslims supposedly massacred cows by the hundreds.[14] These concerns are often explicitly framed in terms of national security. For instance, Gauri Maulekhi has frequently warned of the dangers of cattle smuggling on the grounds that it not only involves the mistreatment of animals but also because of "well-established links" between international terrorism and funds from the smuggling trade.[15] Cow slaughter is thus explicitly represented as a threat to an Indian community and nation that is Hindu in its ethos.

In 2014, when the BJP won national elections in India, demands for an all-India ban on cow slaughter resurfaced in the political arena. Within a month of coming to power, Narendra Modi, the new prime minister, promised that he would press for a national law during his term in power, a law that would put an end to the "pink revolution" (referring to the color of a slaughtered animal's flesh) initiated by the Congress. It was clear even in the lead up to the elections that cows would be making their return to the national stage. Slogans such as "BJP ka sandesh: Bachegi gay, bachega desh" (The BJP's message: Save the cow, save the country) and "Modi ko matdaan, Gau ko jeevdan" (A vote for Modi, the gift of life to a cow) appeared on billboards, in newspapers, and in WhatsApp messages that reached millions of people. Images of Modi bowing his head in front of a cow went viral. The twinning of cow protection and national honor in these slogans and images was difficult to miss. A few months before the elections, a bumper sticker that appeared on cars in Delhi and other cities across the country reminded Indians that the election was an opportunity not just to decide between parties but more fundamentally to choose the kind of nation that they wanted. *Kisey chunenge?* (Which will you choose?), the sticker demanded; *Gau-raksha ya gay-raksha* (Cow protection or gay protection).[16] The sticker warned of the fate that awaited India, and more specifically its Hindus, if the BJP was not voted into power: those who chose *gay raksha* (and, by implication, a party other than the BJP) would be choosing citizenship in an effete state whose purity and power would be diluted by the protection of queer citizens. Those who picked *gau-raksha*, on the other hand, would be putting their faith in a resurgent, masculine Hindu nation united around the symbol of the *gau-mata*.[17]

Following the elections, the objective of cow protection was pursued through both legislative processes in several Indian states, from Maharashtra to Haryana, and everyday acts of violence against anyone who was deemed a threat to the cow. Even a whisper that someone had slaughtered and/or consumed a cow brought violence down on them. In 2015, a mob on the outskirts of Delhi lynched a Muslim man who was suspected of eating beef. A few months later, a Muslim member of the BJP and his entire family were imprisoned under the National Security Act for consuming beef. Vigilante groups of *gaurakshaks*, mostly young men, policed the streets in search of offenders and were quick to dispense "justice" if they found anyone they suspected of cow slaughter.[18] Violence was not visited on Muslims alone. Dalit groups, who have traditionally worked in slaughterhouses and tanneries, were prevented, often by violent means, from continuing in these occupations and also from consuming beef. In 2015, a group of *gaurakshaks* in the state of Gujarat filmed themselves publicly flogging four Dalit men whom they found skinning a dead cow.[19] In 2016, Pehlu Khan, a Muslim dairy farmer, was murdered on suspicion of cow slaughter in Rajasthan. While these acts of violence in the name of the cow have a long history in postcolonial India (Chigateri 2010; Jodhka and Dhar 2003), they took on new, often fatal, intensity in the months after the 2014 election. Indeed, the lynchings have continued unabated well into 2017.

The idea that the cow represents a "key symbol" (Ortner 1973) of Hinduism and of a masculine Hindu nation in particular emerged forcefully in the nineteenth century. It was in this period that cow protection emerged as the cornerstone of an "incipient Hindu nationalism" (Van der Veer 1994). It was now difficult for Muslims and Christians to claim any affinity with the cow; a figure like Ghazi Miyan, the eleventh-century Muslim warrior-saint who was heroized as a "protector of cows and cowherds," had little place in the resurgent Hindu nation that religious nationalists imagined would be united by the figure of the cow mother (Amin 2016).[20] In 1881, Dayanand Saraswati published the *Gaukarunanidhi* (Ocean of compassion for the cow), which was to become one of the foundational texts of the cow-protection movement. In it, Saraswati declared that the slaughter of cows was an act of violence against all Hindus (Van der Veer 1994). At the same time, perhaps recognizing, like Togadia more than a century later, that this argument would not be enough to win popular support for cow protection, he located the need for protecting cows in their economic utility (Adcock 2010). In any case, Saraswati's word was carried across north India in printed pamphlets and through word of mouth by a network of traveling preachers. It spread like wildfire, and *gaurak-shini sabhas* (cow-protection societies) soon started springing up in big and small towns alike. The cow-protection movement soon gained widespread following across northern India, drawing in Hindus of both reformist and orthodox persuasions, wealthy and upper-caste elites, and select lower-caste groups in rural areas into its fold (Pandey 1983).

The Arya Samaj, the organization established by Dayanand Saraswati that was to play an important part in popularizing the program of cow protection, was making quick inroads into the hill districts of Kumaon in this period. Pictures of Dayanand Saraswati with a cow had become quite popular in the town of Almora (Pande 2015). In 1893, the *Almora Akhbar*, a local newspaper, noted the importance of cows to local Hindus. "No improvement in the relations between Hindus and Muslims," it said, "appears to be possible until cow killing has been entirely stopped and when the authorities show no favor to one community over another" (Pande 2015, 37). The "cow-killing issue," as it came to be known, was clearly inflaming passions in Kumaon.

In 1888, the High Court in the town of Allahabad in the United Provinces, the colonial province that included the hill districts of what is now Uttarakhand, overturned the ruling of a lower court that convicted some Muslim butchers for the offence of "destroying an object held sacred" by Hindus.[21] In its reversal of this decision, the High Court declared that Muslims who slaughtered cows could not be charged with insulting the religious sentiments of Hindus because the cow did not constitute an "object held sacred" (Van der Veer 1994).[22] The court's decision provoked outrage and set off a series of violent riots between Hindus and Muslims that have been studied in some detail by historians.[23]

Scholarship on the riots has drawn a complex portrait of the motivations that drew lower-caste groups into the cow-protection movement of the nineteenth century. These riots have been read as providing an opportunity for the "expression of non-elite identity and mobilization" (Gould 2012). For lower castes, it has been suggested, participation in these riots was a chance to assert a higher-caste status in relation to a sacred symbol of upper-caste Hindus, the cow (Pandey 1983). The reasons behind elite support for cow protection, however, are less well fleshed out, with scholars imputing either purely political or religious motivations to upper castes who joined these movements.

The political argument holds that the "importance of religious community in political mobilization for Indian elites to a great extent . . . revolved around the mobilisation of resources for the winning of local power in representative institutions" (Gould 2012, 50). One of the difficulties with such an argument, even as it highlights the importance of the political context in which cow protection came to life, is that it overlooks the religious appeal that the symbol of the mother cow held for a number of Hindus, including the elites who organized these movements. Religion is subsumed into politics in a way that does not allow for an examination of questions of affective power and belief.[24]

The religious argument constitutes the other extreme, with scholars arguing that cow protection was a largely religious issue for upper castes. For instance, Anand Yang argues that "the issue of the cow touched the religious sensibilities of many Hindus deeply because it centered on their fundamental symbol" (Yang 1980: 580). For Hindus, thus, the cow was always a sacred religious matter. Similarly, the historian Sandria Freitag (1980, 606) notes that "the cow was a powerful symbol to call into play because it was sacred in itself."

There are several problems with this line of argument. In particular, I want to reflect on three of them. First, such perspectives take the sacredness of the cow as a given for Hindus. Freitag, for instance, argues that only the symbol of the cow could bridge factional disputes between different Hindu social groups because her sacredness was unquestionable. In these accounts, the cow figures as a timeless sacred sign for Hinduism at large. She is, in effect, a pure and essential signifier of Hindu identity. This argument has unintended but troubling resonance with the romantic Orientalism that represented India, especially Hindu India, as a deeply religious society with "self-born, resilient" symbols (Hansen 1999, 10). In emphasizing the efficacy of the cow as a religious symbol for Hindus, this characterization tends to replicate assumptions crucial to the colonial encounter, such as the idea that all Indians are innately religious (Adcock 2010).[25]

A second and related criticism of this explanation for the rise of cow protection is that it does not adequately consider the conditions in which certain symbols *become* sacred. What makes people willing to shed their own and others' blood for the cow? How does the cow come to be sacred to some Hindus in particular spatial and temporal contexts? How do we understand the undeniable symbolic significance of the cow for Hindus without reducing her to a primordial symbol? Historians and anthropologists have offered a variety of responses to these questions.[26] The anthropologist Marvin Harris famously tried to analyze the taboo on cow slaughter in India within the framework of cultural ecology (1966). For Harris, the sanctity of the cow could be explained as a rational taboo, a cultural mechanism aimed at protecting a low-energy, small-scale, animal-based ecosystem. He focused on the traction provided by bulls and the use of manure as cooking fuel to argue that the sacredness of the cow served its function well. The empirical grounds of this argument were disputed hotly by Harris's contemporaries. However, even its conceptual framework is untenable in its reduction of religious practice to a mere response to ecological necessity.

The historian Peter van der Veer provides a thoughtful and more convincing response to this question of how the cow comes to be sacred. He notes (Van der Veer 1994, 87) that the "sanctity of the cow's body and the prohibition against killing and eating her is *made real* for Hindus in crucial ritual performances that communicate a variety of cosmological constructs." Certain Brahminical rituals concerning death, pollution, and sin, he argues, represent the cow as nourisher, the mother of life, and a symbol of wealth. For instance, in the Vaitarani ritual, which is an important part of Kumaoni religious life, a person who is about to die grasps the tail of a cow with both hands. It is believed that only a cow can help a person in the frightening transition that they make immediately upon their death, when they have to cross the foul river of death to arrive at the opposite bank on which the kingdom of the dead is found. Embodied rituals such as this one make the sacredness of cows visible and meaningful to Kumaoni Hindus. What is valuable about Van der Veer's approach is his insistence that the symbolism of cows must be sought in tangible rituals. His work suggests that the symbolism of the cow is made meaningful through the involvement of the bodies of actual cows. The belief that the cow is a protective mother figure, for instance, is stimulated by the consumption and use of cows' milk. The materiality of the substance thus plays a crucial role in its symbolic enactment.

The importance of bovine bodies to ritual practice, however, raises a question that is most pertinent in the lives of mountain villagers. What happens when the very body of the cow that is so ritually and symbolically significant is transformed? A cow is not simply a cow in this economic and affective landscape. She is a Jersey, a *pahari*, a *dogalla* (hybrid). Her body is not an inert receptacle of symbolic meaning but is lively and unpredictable in ways that make it difficult to contain. To put it differently, her body is not just constituted by ritual practice but is *generative* of it. When one kind of cow replaces another in this complex of economic and symbolic connection, she exposes and shifts the grounds of relatedness.

Business Cows

Soon after coming to power in 2007, the BJP not only passed the ban on cow slaughter but also announced their intention to revitalize dairy development in the state, a project that had already commenced under the previous Congress government. In keeping with the broad Hindu nationalist emphasis on the economic utility of the cow, the government declared that the cow would become the motor of rural development. One of its primary goals was an increase in milk production that would unleash a "white revolution" along the lines of what the Amul cooperative had accomplished in the state of Gujarat.[27] To realize this increase in milk production, the state increased its efforts to extend breeding services in mountain areas with the intention of "upgrading" local varieties of cattle through artificial insemination (AI). Jersey semen proved to be most popular with mountain villagers because of the Jersey's flexibility on steep terrain. The hybrid cow born of this cross insemination was referred to by villagers simply as "Jersey," although some hybrids who looked more visibly *pahari* were called *dogalla*. The (hybrid) Jersey quickly began to emerge as the bovine face of the white revolution in the mountains. In 2012, villages in the subdistrict where I conducted fieldwork had 5,566 cows of the "exotic/crossbred variety" and 13,189 cows of the "indigenous" variety.[28] By 2017, the number of crossbred cows was still steadily increasing in the villages where I conducted fieldwork.

   While local varieties yield a little less than two liters a day, cows who are born from this crossbreeding can yield almost double that amount. Indeed, it was the desire for increased milk production that led to the appearance of Jersey cattle in India in the first place. They were brought there by British colonists who were keen to encourage dairy production on a large scale; most indigenous breeds of cattle were not productive enough to permit an expansion of the scale of milk production. Milk was thus "caught up in the geopolitical territorialisation of the colony" (Saha 2016).

   In the postcolony, too, milk remains implicated in projects of control over territory, community, and bodies—especially of cattle. To increase milk production one must tinker with the bodies of those who produce it, a recognition that was at the heart of dairy production in Uttarakhand. To achieve greater reach for this project, a growing number of young men from towns and villages were trained by the state as paraveterinarians (paravets), mobile workers who would provide basic breeding and health services to village cattle. These paravets charged villagers anything from 150–200 rupees (US$2–$3 in 2016) for these services. Parallel to these efforts by the state, some NGOs also began to equip young locals with the equipment and start-up capital they needed to deliver "artificial insemination and simple animal health services to their local community."[29] Alongside its focus on breeding and health, the Uttarakhand Livestock Development Board revitalized the existing state dairy cooperative, popularly known as Anchal. During the first term of the BJP government in Uttarakhand, a number of new dairy societies were set up in villages under the patronage of Anchal. The spread of dairies was critical, state officials and agricultural scientists argued, because it had the potential to effect economic transformation in rural Uttarakhand, where almost every household owns some dairy livestock.

The transfusion of the Jersey's "exotic blood" into the indigenous bovine bodies that dominated this region created a cow that was familiar but also materially new. I often heard from villagers that the Jersey was an entirely different cow from the *pahari* in body and behavior. Where Jersey bodies were small and rounded with big heads, *pahari* cows were thinner and had smaller elongated heads with pointy horns. Pahari cows have a pronounced *jatta*, a hump, where their neck meets their back. People noted with wonder and mirth that Jerseys even held their bodies differently from *pahari* cows. Their walk was *matak matak* (swaying heavily from side to side), but *paharis* walked *laccham laccham* (with a gentle sway). Jerseys were gentle and allowed anybody to touch them, while *paharis* were ill tempered and would gore anybody other than their owner who had the temerity to come too close. They were, people often repeated, *ladaku* (quarrelsome) cows.

For many villagers, the state's active role in the (re)production of Jerseys only highlighted the unfamiliarity and difference of these cows. The work of the traveling paravets who could render a cow pregnant with the insertion of just one needle was regarded with some wonder, especially by older villagers, one of whom once described a Jersey calf to me as having been born from the "needle of the state" (*sarkari sui*). Indeed, some paravets themselves saw their work as full of mystery even though they knew the science behind insemination well, having absorbed it during their training. One of the paravets who worked in the villages where I conducted fieldwork described AI as *magical*, a term that captured the enchantment this process held for villagers quite perfectly. Villagers would often ask each other whether a calf was born of an injection or a *saand* (bull), and some were never quite convinced that the

FIGURE 14. *Pahari* cows with their distinctive hump.

semen injected into a cow came from an actual bull. "But it's frozen," one elderly man complained to the paravet. "How does it retain its heat? The state must definitely tamper with it in some ways." No amount of explanation by the paravet would make him believe otherwise. "What kind of cow is this?" he asked. "Our *pahari* cows are only born from the hot semen of a bull. No normal cow can be born from an injection unless it's a special injection of the state."

His friend whose cow was being inseminated interrupted him. "What do you mean 'normal cow'? Look at this cow. She gives seven liters of milk a day. And you should see her *style* when she stands up to be milked. This is the cow of the future. Obviously it's different from our cows." The man who doubted the genuineness of the semen had to nod at the obvious sense of this. The paravet whom I was accompanying winked at me as the old men continued their conversation. On our way to this job, he had told me that many older villagers, even after years of having witnessed numerous births from AI, were still puzzled by it. The man who had declared that the Jersey was the "cow of the future" went on, intent on resolving his friend's misgivings. "There's no problem," he said. "You keep a *pahari* cow for its dung, its urine, and for *puja*, and you keep a Jersey for the milk. One for home and one for the outside" (a phrase commonly used to denote the existence of a wife and a mistress).

The man's characterization of the Jersey as a cow of the future was common among villagers. People called Jerseys "business cows," "modern cows," *style-*

FIGURE 15. Jersey cows.

wal goru (a cow with style), and argued that these attributes were inscribed
onto their very bodies. Women often told me that a Jersey when standing up
to be milked had a different "style," that their stance was unfamiliar and forced
the person milking them to change the position of her own body.

Even the milk that came out of these cows was different. Jersey cow milk,
though more plentiful, was thinner and less strongly flavored than the milk
of *pahari* cows. Their urine and dung, similarly, were watery. These products,
people argued, simply did not possess the same nutritional and health benefits
that the milk, urine, and dung of *pahari* cows did. This belief in the nutritional
superiority of the products of *pahari* cows was reinforced by the opinions of
state officials and local scientists. In 2012, a scientist from the neighboring
mountain state of Himachal Pradesh declared that the milk of *pahari* cows
contained a type of protein, A2, that could help people fight a range of diseases.[30] The milk of "exotic" breeds such as Jerseys, he added, contained a different protein, A1, and therefore did not have the same health benefits. Ganga
Ram, the villager who told me about the article, was triumphant that his belief
in the superiority of the *pahari* cow's milk had been legitimized by science.
Another woman told me that she would apply a *pancagavya* (a mixture of
milk, curd, butter, urine, and dung) made *only* from the products of the *pahari*
cow to her blisters and sores. In 2011, a state official charged with handling
livestock improvement schemes told me privately that even though Jerseys

FIGURE 16. Villagers claimed that Jerseys had a different style of being milked.

produced more milk than *pahari* cows, their milk was not as nutritious and would cause health problems for villagers in the long run. Even their dung, she said, was not as good a fertilizer as that of the *pahari* cow's.

Such beliefs were only reinforced by traveling preachers. In 2016, relatives of the family I lived with in Pokhri told me about a *gau-katha* they had attended at a temple in Almora a few months earlier.[31] The preacher was Gopal Mani Maharaj, the founder of the Bhartiya Gau Kranti Manch, a prominent cow-protection organization that has been at the forefront of agitations demanding that cattle smugglers and slaughterers be given the death penalty for their crime. Along with the rest of the family, I watched a video recording that the eldest daughter-in-law had made of the *katha*. In it, Gopal Mani described the many virtues of *pahari* cows to an enraptured audience composed mostly of women. "Listen to them carefully," he said. "When they call, they say *maaaa* (mother). Foreign cows just say ummm. That's the kind of *shakti* our cows have." The daughter-in-law broke in to say that she thought that *pahari* cows always sounded as if they were blowing a conch at a *puja*. Meanwhile, Gopal Mani was now speaking about the many health benefits that could accrue from keeping *pahari* cows. It was not just their milk that was full of nutrition, he said; even asthma could be cured by breathing in the air released from the mouth of a *pahari* cow.

Ironically, then, a number of Hindutva ideologues are among the first to claim that *desi* (Indian) cows are superior to foreign cows and that the latter will prove to be ultimately destructive for the spiritual and physical health of the Hindu nation.[32] Given that milk is an importance substance of kinship not just in South Asia but in many other contexts, it is no surprise that the quality of milk provided by foreign cows is singled out in these denunciations of them.[33] In 2015, Shankar Lal, a Rashtriya Swayamsevak Sangh (RSS) member, declared that the milk of Jersey cows was full of "poisonous particles" that would make the person who drank it "think impure thoughts and do wrong deeds[, which] results in increase in crimes."[34] Similarly, in Kumaon, Pratap Pant, the head of the local chapter of the Bharatiya Gau Kranti Manch, an organization devoted to protecting cows from slaughter, told me that his organization had requested the government to ensure that children in *anganwadis* (government sponsored childcare centers for children up to the age of six) be provided milk from only *Pahari* cows; this was, he said, to ensure that children are not only physically but also morally healthy. "Our roots as Hindus," he told me, "are in *gai* [cow] and Ganga. If these children drink the milk of *pahari* cows, they will grow up virtuous [*sanskritik*]. Only when they drink this milk will they understand what a great sin [*ghor paap*] cow slaughter is." As I listened to Pant, I was struck by how much he associated the *pahari* cow with moral and physical strength. In his view, it was only the milk of the indigenous mother cow that could birth a spiritually and materially powerful Hindu nation.[35] Her milk binds those who drink it to her as children; they must pay off the debt of her milk, her *mamta*, by vowing to protect her. It is because she feeds the nation her milk—the stuff of kinship—that she must be protected.

And yet it is her milk that is also the source of trouble. The Indian state's push for dairy development has increased the number of cattle not just in Uttarakhand but across the country; as of 2017, India had the largest number of milk cattle in the world. Nonindigenous varieties of cattle have been crucial to sustaining the dairy revolution given that they produce so much more milk than indigenous cows. The massive scale of the "white revolution" that Hindu nationalist leaders, including Narendra Modi, seek to bring about thus depends on the very same bovine bodies whose foreignness they decry. However, this program of economic revitalization is difficult to align with the ban on cow slaughter After all, beef and leather industries in India are direct outcomes of the same dairy industry that has always been a critical feature of postcolonial development in India but even more so in ideological and material terms under BJP governments. As Christophe Jaffrelot (2008) points out, the logical end of the economic rationale for protecting the cow is in the slaughter of old and sick cows that have become useless.[36]

Old and unproductive cows were a major drain on meager household resources, especially labor, in the mountains. Villagers were quick to point out that the success of AI initiatives, which were more likely to produce offspring than traditional breeding, meant that their cowsheds were now full to the point of overcrowding. Young bulls who were of no use for agriculture in villages situated along ridges, were increasingly abandoned and could often be found huddling in small groups by the highway. Finding a local buyer for bulls was so difficult that most farmers would have to pay someone as much as a thousand rupees or offer them a couple of bottles of alcohol to take bulls away. Even though the people who took these bulls promised that they intended to use the animals to plough their own fields, everybody involved knew that the majority of animals would soon be deserted, either in the forest or by the roadside. Several people said that their bulls had returned home a few days after being given away; they speculated that the bulls must have re-traced their steps after being abandoned by the persons who had taken them. Others spoke of having learned only months later that the animals had been deserted. One woman told me indignantly about how she had given her bull, and five hundred rupees, to a man who lived near her village, only to find out from her relatives that he had let the animal loose in the forest on the same day that he took him from her. Animal-rights groups and Hindu nationalists alike bemoaned the fact that cattle smugglers did not even have to pay for animals anymore but could pick the animals up for free on the road. By 2017, when the state's crackdown on slaughterhouses in Uttarakhand and in the neighboring state of Uttar Pradesh had intensified and the threat of vigilante violence loomed large, it became common to see large numbers of stray cattle, mostly bulls but also a few old cows, grazing alongside the highway.[37] Many people complained that cattle smugglers had stopped coming to pick up the animals because of the risk involved, leaving these animals to run amok in nearby villages.

The contradictions between the Hindu right's support for dairy and their opposition to any and all cattle slaughter made it difficult for Hindu ideo-logues to publicly renounce Jersey and other foreign cows entirely. While Pratap Pant and others gau-rakshaks like him in Uttarakhand were convinced of the moral superiority of *desi* (Indian) cows over Jerseys, this recognition did not impede their belief that the slaughter of *any* cow—*desi, videshi* (foreign), or hybrid—within the borders of India was a threat to the *gau-mata* of the Hindu nation. State laws on cow slaughter, too, make no distinction between the different kinds of cows they protect.[38] When I asked Pratap Pant whether he felt that killing a Jersey cow was as great a sin as killing a *pahari* cow, he was emphatic that the "murder" of any cow would lead to the nation's destruction. "As long as cows are slaughtered," he said firmly, "the nation cannot progress. No amount of good work—reading the epics, meditating, praying—will yield fruit. That is what made Gopal Mani Maharaj turn his attention to putting an end to cow slaughter. We believe that killing any cow is a sin, whether that cow is *desi* or *videshi*." Thus, while the leaders of cow-protection organizations might declare that only *desi* cows are worthy of love, care, and veneration, any cow, even a Jersey, becomes a mother cow when it comes to the difficult issue of slaughter. For them, the Dalits and Muslims who kill a cow, any cow, are enemies of the Hindu nation and, by extension, killable. There was no question for people like Gopal Mani Maharaj, Pratap Pant, and other key figures of the cow-protection movement that cattle slaughter should be a crime punishable by death. Indeed, vigilante violence in the name of the cow—whether *desi* or *videshi*—was excused and even encouraged by many of the leading figures in contemporary cow-protection movements in Uttarakhand and across the country.

While one could thus argue that ordinary villagers and cow protectionists both practiced a "nativist biopolitics" in relation to cows, I would caution against a seamless equation of these two nativist imaginaries (Müenster 2017). What distinguished one from the other were not only the contradictions between Hindu nationalist theory and practice or their vigilante violence, but also the terrain on which such distinctions between different cows were drawn and what they meant for people's relationships with these animals. For people such as Gopal Mani Maharaj and Pratap Pant, it was the foreignness of the cow that was the problem; theirs was what Daniel Müenster (2017) calls a "bio-nationalist" critique, rooted in transcendental ideas about the racialized purity of the Hindu nation. For *pahari* villagers, however, the fear was not that consuming Jersey milk would cause a loss of national or moral purity or, indeed, turn them into criminals but that the Jersey's body was not ritually powerful enough for her to be part of everyday regenerative religious and social practice in the same way that *pahari* cows were. Neeta, one of the women with whom I had watched the video of Gopal Mani Maharaj's *katha*, told me she agreed that the milk of *pahari* cows was probably more nutritious than that of Jersey cows. But, she said, a Jersey cow's milk was good too. In particular, she rubbished Gopal Mani Maharaj's claim that drinking the milk of a Jersey cow would eventually destroy one's moral compass. "Jerseys are good cows," she said. "I like my Jerseys very much. They don't have as much *shakti* as our *pahari* cows, but that is because they are not of the mountains. But then even a *desi* cow from the plains has the same problem."

The singularity of the connection that *pahari* cows shared with the place, then, was crucial; many believed that cows indigenous to other parts of India would not be able to substitute for *pahari* cows because they were not an intimate part of the historically *situated* social relations that constituted the *pahar*. I often heard that *desi* cows from the plains could not be an adequate substitute for *pahari* cows in certain rituals because they did not possess the substance of *pahariness*. The salient distinction for many people was the narrow one between *pahari* and non-*pahari*, not the broad one between *desi* and *videshi*. Indeed, some villagers told me that it was possible for a Jersey cow with several generations of ancestors who had lived in the mountains to take on *pahari* characteristics. "If they live in the mountains long enough," one man said, "even their identity (*pehchan*) can change. Our ancestors came from other places, but over centuries we have become *paharis*. That can happen with cows too." Thus, even as people distinguished between different cows and made decisions about life and death based on those distinctions, these categories of difference were relatively fluid and open to change over time.

For villagers, then, what distinguished one bovine from another was their particular embodied history. These differences, people told me, became apparent to them through their affective proximity to these animals, a proximity established through the everyday routines of care and labor that I described in the last chapter. The multispecies rural economy of the mountains was characterized by an intimate bodily knowledge that emerged through practices of what Anna Tsing (2015, 2) calls "collaborative survival." As people, especially women, and cows labored for and nurtured one another day after day for years on end, they came to know each other well. The distinctions people made between different kinds of cows emerged from precisely this intimate understanding of the other, an understanding that was grounded in the recognition that each needed the other to survive and even thrive. For instance, when I asked Rekha *chachi*, a Dalit woman, how she knew with such certainty that the Jersey did not have the same *shakti* as her *pahari* cows, her response was quick. She knew, she said, because she spent the better part of every single day of her life with these animals. She told me how she had learned this difference through the process of repeatedly laboring on a body that became as intimate as her own. Her life was yoked to theirs, and it was this yoking—and the care and love that accompanied it—that was the source of her knowledge. She said,

> I know these animals as well as I know my own children and grandchildren. I'm with them from morning to evening cleaning them, taking them to graze, feeding them, cleaning out their *gobar* (dung), milking them, and just sitting with them. I know when they are hiding milk for their calves, and I know when they stop giving milk because somebody has turned the evil eye upon them. I know their *svabhav* [nature]. That's how I know that they don't have the same *shakti*. Try doing a *parikrama* [a ritual circumambulation] of a Jersey cow. They don't stand still; they just keep moving with you. But our mountain cows

just know that they must stand still until a person has touched their head to all four feet. You get to know the difference between them by living with them.

It was such affective intimacies that were foregrounded in villagers' explanations of how it was they knew that the Jersey was not a ritually powerful cow. While people like Gopal Mani Maharaj shared their ideas about the difference between these animals, their claims were disembedded from a rural economy where caring well for the other was crucial to collaborative survival, even if one of the partners in the relationship was destined for eventual death at the hands of the other. For such preachers, the Jersey cow, even if she could not be slayed, was a creature unworthy of care, consideration, and respect. For women like Neeta and Rekha *chachi*, on the other hand, the Jersey, even if she was killable, was a cow to be nourished and cherished. In this multi-species economy where love and death were often inextricably intertwined, the certainty that one would eventually have to send an animal to her death did not preclude care, gentleness, and affection for her in the ordinary course of life; indeed, as I argued in the last chapter, the knowledge that they would bear direct or indirect responsibility for an animal's killing often led people to cultivate ethical sensibilities within the course of their everyday interactions with these animals. The Jersey's difference from the *pahari* cow might have made her killable as far as many villagers were concerned, but it certainly did not make her unlovable in the way that many right-wing Hindu ideologues and gau-rakshaks imagined.

However, while this recognition of difference did not mean that villagers like Rekha *chachi* cared for and loved their Jerseys any less, it did indicate to them that the Jersey's body was not appropriate for the daily acts of ritual through which, as Van der Veer points out, the sacredness of the cow and her symbolic significance is recreated in local contexts. Let us turn, then, to how people responded to this altered materiality in everyday rituals contexts involving cows.

Ritual Cows

It was the eleventh day after Binduli, a large brown Jersey cow, had given birth to a female calf. Early that morning, right before sunrise, my friend Manju Bhatt bathed and made her way to the cowshed, the empty milk pail she held in her hand clanging in the dawn silence. She milked Binduli with some difficulty, bringing the calf to her udders to trick her into releasing the milk. Filling the pail with milk, she walked back to the house to make ghee (clarified butter).

The village priest would stop by in a few hours to name the calf, and there was a lot to be done before that. She pulled out the electric churner that her husband had bought in Haldwani a few years ago and churned the milk to make butter, from which she then made the ghee. That done, she immersed herself in the task of frying *puris* (fried bread), which had a little drop of milk mixed into the batter, and making *kheer*, a mixture of Binduli's milk with the finest rice in the house. I asked if all the food was to be made with Binduli's milk, and she nodded. "We'll feed her and the calf first, then the priest, and finally, the rest of us will eat. This is the first day that people are allowed to drink the milk. If you drink a mother's milk before eleven days [after the birth of a calf], her teats start releasing blood instead of milk." By eleven, the priest had arrived. He carried the loaded plate of *puja* offerings to the cowshed—a little pat of cow dung; five white stones to represent Airy, the god of livestock; a little earthen lamp, its wick glistening with oil; some *puris* (fried bread), butter, ghee, a mound of lentils, *badis* (fritters), and two *batashas* (sugar sweets); a bronze pot filled with *sindoor* (vermilion); a few flowers; and a glass of milk and one of cow urine. Manju followed him into the dark corner where Binduli was tied. Her calf lay in the fold of her belly, snuggling into her warmth. Manju tied a frond of grass around the calf's neck while the priest set the plate down in front of Binduli. He sprinkled some cow urine on Binduli and her calf, and fed them a little of everything on the plate. A swipe of *sindoor* on their foreheads, the chanting of a short *mantra*, and the naming ceremony was done. The ceremony itself was surprisingly short given how long and elaborate the preparations had been.

The priest left after lunch, and it was finally our turn to eat. After heaping her husband's plate with more food, Manju poured him a glass of milk. He drank it in one gulp and burped with satisfaction. "Whoever buys this milk at the dairy tomorrow will be very happy." There was a deep note of concern in Manju's voice as she asked, "Should we be giving the milk to the dairy tomorrow itself?" "Why not?" Mohan asked nonchalantly. "What if Airy is angry?" Manju inquired. Most villagers believed that the milk of a cow who had just given birth was not to be consumed by anyone outside the household for the first twenty-two days or the *dyavta* Airy would punish the family.

Mohan responded to her fear with a dismissive snort. "Don't be so superstitious. The *dyavtas* care about these rules of purity-pollution (*chhoo-achhoot*) only when the cows are *pahari*. *Par yeh to* hybrid Jersey *gai hai* (but this is a hybrid Jersey cow). They're bred to give four or five liters a day. Nothing will happen if we sell it before twenty-two days."

Manju seemed convinced by this argument. It was true, she told me later, that this cow was unlike their old *pahari* cows. "Our cows have grown up here like us. Just like we are *paharis*, they are also *paharis*. They have breathed this

FIGURE 17. The offerings consisted of a pat of cow dung, five little stones to represent Airy, and an earthen lamp.

air for generations, and their bodies are nourished by our mountain herbs and plants. But Jerseys are new here. They prefer Kapila Pashu Aahar [a particular brand of cattle feed] to eating grass. How can they be the same cow?"

Like my friend Neeta with whom I had watched the clip of the *katha*, for Manju there was little doubt that *pahari* cows were constituted differently than Jersey cows by virtue of their long history of exchange with the sacred geography that they inhabited. As far as she was concerned, *pahari* cows shared a kinship with *pahari* people in that both materialize through a long history of daily exchanges in and with the physical, social, and ritual environment of the mountains. The relatedness between them was enforced through everyday ritual performances in which the consumption and use of the ritually powerful bodily substances—milk, dung, urine—of *pahari* cows connected people to them in intimate ways; through these transactions, people were established as kin not just to *pahari* cows but also to the gods.[39] In other words, the sharing of ritually charged substances was at the heart of this kinship. Jersey cows could not be a fundamental part of this particular relational ensemble of *pahari* people, cows, and gods because their bodily and moral substance was formed through different practices of exchange. And yet, as I mentioned earlier, the possibility that Jerseys could—through exchanges with *pahari* humans, animals, gods, and other natural forces—become *pahari* over time remained open.

Over the course of my time in the mountains, I heard from many others about how their relationship with Jersey cows was different from their relationship with *pahari* cows. The varied nature of this relatedness centered on the differential ritual power of bovine bodies and their products. A common refrain was that Jerseys were good cows but *bekar* (useless) when it came to *puja*. One of my friends once sent me to her neighbor's house to ask for a pat of cow dung from their *pahari* cow. She was doing a *puja* that needed *gobar* and did not want to use the runny and stinky *gobar* from her Jersey cows. Pahari *gobar*, she told me, was a firm and perfectly rounded pat that "smelled good."

On another, more serious, occasion, the lack of *pahari* cows in a household became a real impediment to the *vaitarani* death ritual. In 2011 Bina Bhatt added another cow, this time a *pahari*, to the four Jerseys she had already. The cow was given to her by her maternal family. Bina *mausi*'s mother had died a few days ago, and it was this cow whose tail she had held onto as she passed away. This cow had helped her cross the mythic river of death to the safety of the opposite bank. *Mausi*, who loved this cow, Kauli, from the first day she got her, told me of how much trouble her family had gone to in finding a *pahari* cow for the ritual. "My mother refused to die until they brought her a mountain cow. She only ever kept mountain cows. She always said they were *real* cows. So we looked here and there for a mountain cow. We went mad. But only when we brought her one did my mother finally die." When I asked Bina *mausi* why she thought Jersey cows didn't have the same power as mountain cows, she was quick to respond.

> My sister-in-law tried to convince my mother to hold onto a Jersey's tail by saying that one cow is like another cow. But you can see that these cows are different. We would use the urine and the dung of our mountain cows to purify the house. Their urine had a rich color and their dung was solid. The dung of these Jersey cows is watery. And they behave differently, too. My mother is right. The mountain cows we grew up with had tremendous power. They wouldn't let a woman who was menstruating anywhere near them. Nowadays women tell their daughters and daughters-in-law who are menstruating that they can drink the milk of a Jersey even before the third day of their period. Can you imagine? And nothing happens to these Jerseys. If we had done that with our mountain cows, then blood would have come out of their udders instead of milk. Our cow once fell down and broke her leg after my mother gave a glass of her milk to a menstruating woman. That's the kind of *shakti* they have.

Bina *mausi*'s belief that Jersey cows did not possess *shakti* transformed the nature and performance of a key ritual. For her, the ability of different bodies to relate to one another, to affect one another—the keystone of such

FIGURE 18. The belief that Jerseys lacked *shakti* did not foreclose the emergence of affective intimacies between them and villagers.

rituals—was grounded in a shared connection to the social and material environment of the Himalayas. A Jersey's lack of such a connection was visible in her very body, which was why it was, for Bina *mausi* and others like her, a ritually feeble body.[40] This was why, I was told by many, it was not a *paap* (sin) to sell Jerseys to people who could be buying an old cow only for one reason—slaughter.

And yet, while many people did sell or, increasingly, abandon cows who were too old to give milk, this was not an easy decision. Many villagers I knew expressed profound regret at being forced to part with an animal who had nourished them and their families with her milk. The prospect of the death of an animal in whose life they had been so intimately involved was an occasion often marked by the experience of intense grief and guilt for people, especially when they were complicit in these deaths. Unlike the organic farmers of Kerala who, Daniel Müenster (2017, 32) observes, "sever affective connections with increasingly unloved [foreign] bovine others," the villagers I knew had a deep love for their Jersey cows even if they considered them spiritually effete. To emphasize what I said earlier, while farmers might have discriminated between different cows when it came to death, they did not do so in choosing whom to love.

There is one moment in particular that I want to linger on for how much it revealed about the affective intensity of these ties. On one morning, while I was living with my friend Mamta, a man arrived at her house to buy her Jersey cow, Chanduli. Chanduli had recently stopped giving milk, and her care was proving difficult for Mamta. Chanduli was a rather large individual, and the everyday labor of cutting and carrying home enough grass and oak to feed her and the family's other three cows had left Mamta with a permanently pinched nerve. Mamta was now pregnant, and the doctor had told her to cut back on heavy labor. It made sense to sell Chanduli, although it had taken a few months for the family to convince Mamta, who was devoted to the cows. Mamta's brother-in-law, who lived in Rudrapur, a city in the plains, had negotiated a deal with one of his neighbors, who had agreed to buy Chanduli for six thousand rupees for her; the price was on the lower side, but given Mamta's situation and Chanduli's age, the family decided that they could do no better.

When Mamta asked what he was planning to do with Chanduli, the man was noncommittal and said he had to leave. After he had coaxed Chanduli into the waiting Tempo and shut the door, she began to bellow and tried to kick her way out of the vehicle. Her cries grew louder as the Tempo started to drive away. As the sounds receded into the distance, I felt tears roll down my cheeks. I often accompanied Mamta to the grazing grounds, and Chanduli was my favorite of her cows. She constantly poked her velvety nose into our bags in search of food and would look unabashed when scolded for her greed. She loved to have the fold of skin under her neck scratched and would sometimes come up to me and stick her neck out as if demanding a rub. She was the only of Mamta's cows who had not kicked me when I tried to milk her.

When I looked over at Mamta as the Tempo drove away, I saw that her body was racked by great sobs. I reached out to squeeze her shoulder. She touched my fingers with her own in acknowledgment of the touch. "I cry each time," she said, her voice breaking. "Even when the goats are given as *puja*. I spend so much time with these animals. They become part of the family. Love (*moh-maya*) happens." I nodded and told her about how I had wept for days when my family had to put the dog I had grown up with to sleep. The talk of death turned her silent for moment. "I think that man is going to take Chanduli to a butcher," she said finally. Her face was streaked with tears.

"At least she had a good life," I said, trying to comfort her. "Yes," Mamta conceded. After a few moments she added, "The world has changed a lot. Forget sending a cow to the butcher . . . my *nani* (maternal grandmother) said that anybody who hurt a cow by throwing a stone at her had committed a great sin (*thul paap*)." I asked if she thought that she had committed a sin by sending Chanduli off with that man. She thought hard about it.

The world has changed a lot. My *nani* never kept a Jersey cow in her life. If she had, she would know that this cow is very different from our other cows. I loved Chanduli [*accha manti thi*]. I will always remember her. But we never used her dung on any *shubh* [auspicious] occasion. For that we always used Meenu's [their *pahari* cow] dung and urine. I think even my *nani* would have said that there is no sin in letting go of a cow like this as long as you served her well. But still there is sorrow [*dukh*].

Mamta's thoughtful response to my question laid bare the multifaceted nature of people's relationships with the cows that were being smuggled from the mountains to the plains for slaughter. Parting with Chanduli had brought her great sorrow and, over the next few days, it became clear that she missed her immensely. Every time we took the cows out to graze, Mamta would recall how Chanduli had liked to scratch against one pear tree in particular. She pointed out the scrapes and dents in the bark, laughing as she remembered how angry the owner had been when all the pears fell off one summer afternoon after a particularly vigorous itch session. She recalled all the times that Chanduli had wandered into somebody's house in search of food and received a beating for her efforts. Through these acts of mourning, Mamta extended the relatedness she had shared with Chanduli in the past into the present when their life together had come to an end. However, while Mamta often felt remorse and ambivalence about her part in Chanduli's (assumed) death, ultimately she did not regret the decision to sell her. All she said was that Chanduli had lived and served her human family well and that she herself had served and loved Chanduli well in turn. For her, the question of ethics was one that could not be legislated transcendentally but only in relation to everyday matters of care and attachment. One thing was clear to her, however, and that was the fact that all cows were not the same and, therefore, neither was their *shakti*.

"The Cow Herself Has Changed"

In 2012, while I was visiting the family of Puran Ram, we got talking about cows. Their neighbor's cow had died a few days ago. She had been browsing by herself and had suddenly collapsed onto the road under the field she was in. People said it was a snakebite. "She was a Jersey," Puran *da* noted. "A *pahari* cow would have known there was a snake there. Those cows would never have been killed by a snake." The conversation then turned to a description of the difference between Jersey and *pahari* cows. When I asked what he thought of Jerseys, he smiled. They were "modern" cows, he said, and good milk givers even if they did not have the same *shakti* as *pahari* cows. "But for us," he continued, choosing not to specify the collective even though I soon realized he was talking about Dalit villagers, "the coming of the Jersey has led to a big change." When I asked him what the change was, he said, "earlier the Pundits and Thakurs would not let us buy milk from them. They believed their cow would stop giving milk if it was consumed by a Harijan. It was a matter of *chhoo-achhoot.* . . . One night I wanted some milk for my guests to have tea. I asked my neighbor. But they said they didn't have any even though I knew they did. You must know why they said no. It was a matter of great shame for me."

Many upper castes believed that a cow would be ritually despoiled (*bigad jaana*) if a lower-caste person drank her milk. In several kitchens in upper-caste homes, I had observed a packet of *Anchal* milk bought from the dairy or local store kept next to stainless steel vessels full of boiled milk from their cow. When I asked these villagers why they kept these packets of milk, they would say it was in case a low-caste person stopped by and wanted a glass of tea. The Dalit villagers I knew deeply resented these forms of caste "humiliation" (Guru 2009). Puran *da* had told me on several occasions that he would rather not drink tea in upper-caste households than suffer the humiliation of being served Anchal milk from a packet. "But now," he said, "many upper-caste villagers serve us milk as long as it is from their Jersey cow."

"What changed?" I asked. "Arre, ab to gay hi badal gayee hai" (now the cow herself has changed), he cackled. "These Jerseys don't have the same *shakti* as our mountain cows." I nodded at this familiar refrain. "Think of this," he continued. "With all this milk, people send so many liters to the dairy. Anybody could be drinking her milk after buying it at these dairies. Even Muslims might be drinking it. So, if Muslims can drink the milk without the cow falling ill, what [kind of threat] are we [to upper castes]?" He laughed. "When the cow changed, everything changed. Now just look at how many people around us are sending these cows to the butcher. Would they have dared to do that with a *pahari* cow? Their whole family would have been destroyed by the wrath of the *dyavtas*. But there's no problem with such a cow. The relationship with them is just not the same."

What would it mean to take Puran *da*'s observation—that the cow herself has changed, thereby changing the forms of relatedness that are rooted in bodily exchanges with her—seriously? His pithy remark illuminates why Hindu nationalist projects of cow protection had faced such obstacles in putting a complete end to smuggling and slaughter in the decade immediately following the ban: the discourse of cow protection—even as it recognizes the difference between different cows—ultimately seeks to embed an abstract and homogeneous symbol of the cow mother in the bodies of cows who are neither abstract nor homogeneous to the people who live alongside them. The distinctiveness of different bovine bodies matters a great deal to those villagers who are called on, in the law and in the discourse of *gau-rakshaks*, to unquestioningly accept kinship with cows at large. For them, to be related to a *pahari* cow is a materially, symbolically, and affectively singular experience that cannot be replicated in the connection they share with Jersey cows. I can't help but think here of Bina *mausi*'s mother who refused to die until her family brought her a *pahari* cow, whose body she believed was the only one capable of helping her make the transition to the other world; her enduring trust in the *shakti* of *pahari goru* points to how the distinctive materialization of different bodies shapes their symbolization. It reveals how the liveliness of individual bodies shifts how people come to relate to them. In other words, what people like her and Puran *da* were insisting on was the fact that animals condition political and cultural possibilities not just as immaterial metaphors but as particular actors with complex lives, histories, and characters. I take their insight as the point of departure for the next chapter, in which I examine how contemporary politics of *pahari* belonging are transformed by the arrival of monkeys from the city whose distinct histories are mapped on their body in ways that emphasize their stark difference from local monkeys.

Outsider Monkey, Insider Monkey
On the Politics of Exclusion and Belonging

I first heard about the "outsider" (*baharwale*) monkeys within a few days of starting my fieldwork in 2010. I was in a shared taxi, returning from the town of Haldwani to Pokhri, when the driver casually mentioned that at least a hundred monkeys had been dropped off a couple of weeks earlier in the village of Harkoli, a hamlet of about fifty or so houses perched precariously on a hillside right below a highway. A little after midnight, those closest to the road heard a truck pull to a stop in the little patch of oak and pine forest that adjoined the village. And then commenced a *halla* (hullabaloo). The noises that drowned out the gentle drone of the familiar creatures of the night were so terrifying that no one stepped out of their homes to investigate what was going on. The next morning, the monkeys were everywhere. They were too many to be chased off. Even the dogs who bounded out of their homes, tails stiff with joy at the sight of these new adversaries to be harried, returned in a trice, some with injuries from the encounter. Within the week the monkeys had destroyed the entire plum harvest, causing thousands of rupees worth of damage.

The other occupants of the car clucked in sympathy at the plight of the *Harkoliwallahs*. One woman said with some bitterness that her village had been dealing with these "outsider" monkeys for the last few months, and she knew how destructive they could be; they had eaten everything in sight—cucumbers, squash, beans, apricots, peaches, pears, and plums. There was no getting rid of them. Soon the others in the car began to name villages that were similarly under siege by monkeys from elsewhere; the list seemed endless. In many places, I was told, people had given up cultivation altogether because the monkeys would eat everything and their labor would come to naught. *Pahari* monkeys raided crops too, I was told, but the scale and intensity of the depredation by these new monkeys was a tipping point. A man in the car recounted

FIGURE 19. The outsider monkeys caused particular damage to the fruit crop.

to us how his family had hired the son of a Nepali laborer to spend the day patrolling their orchards with a slingshot; however, the boy had been so badly bitten by the monkeys that his father refused to send him to work anymore even if it meant losing the money that had extended the family income.

Those who talked about the monkeys always echoed a few common themes. "They have come from elsewhere" was a common refrain, as was "they were not here earlier." When I asked how and why the monkeys had come there, "there" being a particular village or the *pahar* in general, the response was always "unko yahan chhod diya gaya hai" (they have been left here) or "truck me bhar-bhar ke aate hain" (they came by the truckload). When I asked where the monkeys were from originally, some responded that they had come all the way from Delhi, Vrindavan, and Mathura.[1] Others declared that the monkeys were exiles from the cities of Haldwani, Nainital, and Almora, banished from these urban spaces at the will of their influential human residents and dumped on powerless villagers out in the middle of nowhere. There were also a few dark mutterings that the monkeys had come from Dehradun, a political ploy to ensure that Garhwal would prosper while Kumaon would be ravaged by monkeys. Their exact provenance was a matter of some mystery. But on this, at least, everyone was agreed—the monkeys had come from elsewhere. And their numbers were only growing, *din ke din, raat ke raat* (day by day, night by night). In 2007, the Congress Party's manifesto promised to tackle the "monkey menace" if voted back into power in the state. The president of the Uttarakhand

Congress Committee, Harish Rawat, explained that the issue had made its way into the manifesto because farmers in every village they visited had begged them to do something about the growing depredation of wild creatures.

However, it was not simply the pillage of their crops and houses by monkeys that so enraged villagers in Kumaon. Predation by monkeys, after all, is a timeworn element of rural life, and conflict is only one dimension of a complex human-macaque relationship.[2] As one older woman put it, it was in the nature of monkeys to pilfer—they were inveterate *chor* (thieves), just like humans.[3] Unlike the "thieving sparrows" of the Cumbum Valley (Pandian 2009), their acts of robbery were believed to be driven more by a base desire for entertainment than by real need. Monkeys would not be monkeys if they did not steal, one man told me; it was a characteristic of their kind (*unki jaat hi aise hoti hai*). People were thus accustomed to the pilfering tendencies of monkeys, born supposedly from their innate love of mischief. The image of monkeys as mostly good-natured rogues was reinforced in local stories, myths, and proverbs that drew from a broad Hindu religious imaginary; it was these narratives that people conjured up to excuse and historicize monkeys' proclivity for theft and mischief.[4] There was, of course, the frequent invocation of Hanuman, Rama's monkey follower, companion, and protector. But there were also a host of other metaphors and analogies. One woman told me that monkeys were as mischievous as young Kanha (Krishna), a lovable thief who could not resist stealing butter.[5] An older man who was famous for his mastery over the *Ramayana* compared monkeys to the figure of Sugriva, the deposed king of monkeys, who lived a life of debauchery until his rescue by Rama.[6] "For centuries," the old man told me on one sunny afternoon after having declared his intention to redress my deplorable lack of familiarity with the intricacies of the Ramayana, "monkeys have been *ayaash* (voluptuary) creatures. Only *bhagwan* Rama can reform them. That's why you have to forgive whatever they do; it's in their blood."

If such plunder of agrarian and household produce was a familiar and forgivable offence, why then did human conflict with monkeys become such an incendiary issue in the years following the creation of Uttarakhand as a separate state? Coding these encounters between humans and monkeys as yet another instance of human-wildlife conflict—that is, as tussles over space and food that affect lives and livelihoods—obfuscates the fact that people were relatively tolerant of the infractions of certain monkeys and not of others. It was thefts committed by *pahari* monkeys that could be endured, sometimes with affection but more often with barely restrained wrath. However, the trespasses of outsider monkeys were deemed by many to be intolerable, especially because they were so aggressive. "There is a limit," one woman complained.

Our monkeys are ours, after all, but why should we have to endure the terror of these monkeys who have come from outside? So many people have been bitten that it is hard to step out on the streets without a stick. Even our dogs are terrorized. My dog doesn't go into the garden anymore after his face was bitten by a monkey. He relieves himself near the house. The monkeys who would come occasionally to our fields would never attack dogs like this. But these monkeys can kill a dog.

She was only one of many who shared the sentiment that it was one thing for humans (and dogs) to suffer the depredations of *pahari* monkeys but entirely another to have one's homes and fields plundered by monkeys from elsewhere who responded to people in unexpected and dangerous ways. Some people complained that *pahari* rhesus monkeys would be unable to leave the forest as long as these outsiders occupied villages.[7] "There used to be an *akela bandar* (single male monkey) who would come to our fields occasionally and eat the crops," one woman told me.[8] "But now that these ruffians are here, he doesn't come anymore. Our situation is the same as his," she said bitterly. "Even we will soon have to stay in the forest. This happens to all *paharis*, we are always exploited." In effect, she was claiming that *pahari* monkeys and humans were related by this common history of displacement and exploitation, a sentiment that was echoed by many others who expressed their sympathy for the local monkeys who had been pushed out by these outsiders. People's resentment was thus reserved for the newcomers; it was they who were denounced as unwelcome interlopers. In other words, I am suggesting that people's bitterness and anger was provoked not by simian crop raiding in general but by *which* monkeys in particular were thieving from fields and homes and attacking people. They framed this conflict as a particular battle between *paharis*—both human and macaque—and outsiders, not just between humans and monkeys as distinct species.

As such, the moral panic bore striking similarity to tropes about invasive or nonnative species elsewhere in the world, whether foreign flora in South Africa (Comaroff and Comaroff 2001) or turtles and frogs imported for food to the United States (Kim 2015). What these different cases share in common is, to borrow from Jean and John Comaroff (2001, 628), the deployment of the nonnative species as "alibi, as a fertile allegory for rendering some people and objects strange, thereby to authenticate the limits of the ('natural') order of things." In Kumaon, too, the discourse around outsider monkeys imagined and enacted a politics of differentiation and exclusion. The very term *bahar ke bandar* (outsider monkeys) reflected a widespread rhetoric that was employed to demand and authorize martial action against outsiders. This was a strident

FIGURE 20. The intrusions of *pahari* forest macaques were met with greater tolerance.

and exclusionary politics that sought to harden lines of difference between insider and outsider, native and trespasser.

Yet what also distinguished the use of nativist tropes in this context from moral panics in places such as South Africa and the United States was the fact that it was through this politics of belonging that Kumaoni villagers exposed and critiqued their ongoing history of exclusion. The preoccupation with outsider monkeys reveals how they became powerful material metaphors for anxieties about the displacement and expropriation of *paharis* by people from the lowlands. Highlighting the unexpected presence of monkey outsiders in this landscape opened up a space for *paharis* to debate not only who belongs in the mountains but also, and perhaps more importantly, to whom the mountains belong. By representing these monkeys from elsewhere as rapacious freeloaders who did not play by the rules of the *pahar*, villagers produced a critical commentary on the growing presence and power of human outsiders. At the heart of the matter of the monkeys, then, was the difficult question of who should have material and moral access to scarce resources in a state that was created to address a widespread sense of inequality and neglect among *paharis* in the first place. Having said that, victimhood is a complicated terrain upon which to base political action. As Laura Jeffery and Matei Candea (2006) note,
enacts a politics that is ambiguous. Discourses of *pahari* vic-
emerged around the issue of the monkey menace naturalized
belonging that were dependent on the exclusion and othering

of human and animal others. And yet it was through this morally complicated "politics of victimhood" that people felt empowered and justified to make claims on a state that was widely perceived to have abandoned the pursuit of welfare for its mountain citizens.

In what follows, then, I trace how this nativist politics around monkeys took shape in the crucible of wider societal anxieties about the loss of *pahari* cultural identity, brought on by a widespread sense that the "mountain state" in which *paharis* were supposed to flourish had failed to live up to its promise. As *pahari* youth continued to migrate from the region in ever-growing numbers due to lack of employment and opportunity in the mountains, those who remained blamed their plight on an actively neglectful state that they believed was more committed to securing the interests of rich plainspeople over its own mountain citizens. While villagers were critical of corporate control over natural resources, it was the purchase of agrarian land by outsiders for second homes, hotels, or just speculation that was seen as a particular threat to *pahari* social regeneration.[9] Even those who had sold their own land to outsiders worried that the *pahar* would soon be denuded of all *paharis*, forced out by the loss of their only stable asset—land. Racked by this anxiety, many *paharis* became increasingly suspicious of outsiders and read all kinds of possibilities and futures into their presence. It was in this atmosphere of fear that rumors about outsider monkeys acquired such a sharp edge. A number of villagers speculated that these monkeys were captured in the plains and released into the mountains by real-estate developers and urban land speculators with the intention of forcing mountain villagers off their land by making it impossible for them to continue cultivation. In this narrative, outsider monkeys were only the latest agents of *pahari* cultural erosion.

But what was it about these monkeys in particular that made them such potent signifiers of alienness? It is difficult, at first glance, to understand why translocated monkeys became such potent metaphors for this larger political and affective anxiety. Rhesus macaques are native to this region even if the individuals who were translocated to these villages were not. As such, the presence of these outsiders did not provoke any public or scientific fears about the destruction of native flora and fauna by an invasive species. There is a stark distinction, then, between this situation and other contemporary debates about nonnative species in which what Claire Jean-Kim (2015) calls "an optic of ecological harm" is deployed by activists and state conservation agencies to represent native species as intrinsically valuable autochthons in need of scientific protection from "invasives." Conservationists and animal-rights activists who criticized the translocation of macaques did so on ethical rather than ecological or economic grounds. Indeed, in a striking inversion of concern, they sought an end to translocation on the basis that the experience of being captured and moved to an unfamiliar place was traumatic for the "invasive" monkeys. It was difficult to make a case against the interlopers because they caused a different *kind* of damage to native flora and fauna than *pahari* monkeys even if the scale of their depredations was new. Why then, to return to the question I asked above, did outsider monkeys become the face of this "naturalisation" (Comaroff and Comaroff 2001) of politics instead of, say, *lantana*, a recognized "alien" that has been blamed by ecologists for fueling the routine and uncontrollable forest fires that cause irreparable damage to the flora and fauna of the region?

To answer this question requires understanding how the behavior of the outsider monkeys themselves shaped the nature of this politics of belonging. Even though these monkeys belonged to a familiar species, they were unfamiliar as *individuals* to villagers. When I asked people how they could be certain that these monkeys were "outsiders," they pointed to the monkeys themselves as the source of their knowledge. One man put it rather evocatively when he told me that these monkeys bore veritable maps on their bodies. "We don't need to hear or see the trucks," he said dismissively, when I asked him if he had ever witnessed the monkeys being dropped off. "Just look at the monkeys. Unki chaal-dhaal saaf bataati hai ki yeh kahin bahar se aaye hain [their comportment is a dead giveaway that they have come from outside]." Many others echoed his conviction over the course of my fieldwork. For them, the distinctive demeanor, attachments, and inclinations of these monkeys made it abundantly clear that they were a different category of beings from the more familiar *pahari* monkeys. These embodied distinctions, gleaned through intimate cohabitation, were then harnessed and put to work by the *pahari* politics of belonging based on a sense of victimhood that I will describe over the rest of the chapter. It was clear that to people, the outsider monkeys were distinctive entities whose unexpected presence in the landscape could serve as a powerful emblem for the growing expropriation of *paharis* at the hands of outsiders. With this insight in mind, the following pages move back and forth on an affective terrain where the monkey as signifier and flesh-and-blood monkeys are embedded in and inseparable from one other.

City-Slicker Monkeys

North Indian cities such as Delhi, Jaipur, Mathura, and many others are home to large groups of rhesus macaques (*Macaca mulatta*), called *laal bandar* (red monkey) in Hindi. The population of these macaques has grown substantially since India banned their export for biomedical experiments in 1978.[10] Rhesus monkeys are daring, intelligent, and curious animals; one primatologist attributes their success as a species to what he calls their Macachiavellian intelligence (Maestripieri 2007). They can adapt to and thrive in a variety of circumstances, whether by crop raiding in rural areas or inserting themselves into temple contexts where they are regularly fed by pilgrims. Indeed, some primatologists suggest that "moderate amounts of human disturbance can result in *increased* food availability for monkeys and *increases* in primate population size and density" (Bishop et al. 1981, 155). Living alongside humans, in other words, can often be quite beneficial to monkeys. Anindya Sinha (2005) argues that the "phenotypic flexibility" of macaques and their "behavioral inheritance"—the ability to learn new behaviors by observing other individuals—have been crucial to their success in a variety of habitats. Rhesus macaques, in particular, have been remarkably successful at surviving even in areas with dense human habitation, unlike many other primate species.

Their presence in spaces across the city is simultaneously entertaining, beneficial, disruptive, and threatening. As Ajay Gandhi (2012) notes, urban monkeys are ambiguous creatures—"annoying menace and sublime force, liberal subject and technocratic problem . . . a source of amusement."[11] When they create havoc in India's Parliament or in the Secretariat where matters of the state are deliberated, they are hailed by the general populace as heroes. But on the occasions that they turn their attention to bustling urban settlements— stealing food, clothes, and household objects, and scaring and biting people, sometimes with fatal consequences—their actions are no longer a cause of hilarity. Even those who believed monkeys were a form of Hanuman, and fed them religiously every Tuesday and Saturday, complained bitterly about the ravages wrought by these animals and appealed to city authorities to have them removed from their neighborhoods. As complaints became more frequent and strident, municipal authorities across cities were forced to act. In Delhi, the death of a deputy mayor who fell off his balcony while trying to fend off a group of rhesus was one of the incidents that forced civic authorities to find a solution to what was popularly known as the "monkey menace."[12] The answer was translocation, the capture of individuals from one site and their relocation to another, more suitable place. From the 1980s onward, the capture and translocation of monkeys was increasingly hailed as the most effective and nonviolent method of controlling monkey populations. In the years to follow, there were several successful instances of translocation.[13] For instance, macaques trapped in Vrindavan in 1997 by the NGO Vatavaran, headed by the primatologist Iqbal Malik, were successfully moved to new forest sites that did not have a resident population of macaques (Imam, Yahya, and Malik

2002). In Delhi, too, translocation came to be favored as the most effective solution to the city's simian woes.

Monkey catchers employed by the Municipal Corporation of Delhi (MCD) and the New Delhi Municipal Corporation (NDMC) were assigned to capture monkeys from neighborhoods where their population had swelled. But the problem was where to send these monkeys. In the early years of the twenty-first century, following a Supreme Court order, Delhi had sent captured monkeys to other states such as Madhya Pradesh and Uttar Pradesh.[14] In 2006, the Supreme Court once again ordered the government of Madhya Pradesh to accept more monkeys from Delhi. However, Madhya Pradesh later refused to accept these monkeys.[15] They claimed that the new arrivals were not only putting pressure on forest habitat but were also spreading diseases they had acquired in the city to other monkeys. Eventually, in 2007, the Delhi High Court directed municipal authorities to release three hundred captured monkeys into the Asola and Bhatti Wildlife Sanctuaries, created on land carved out from a former mine. Even though the monkeys released into Asola were fed daily by the Delhi forest department, which was paid to do so by the NDMC, they continued to escape their new home, scaling a wall around the sanctuary and attacking nearby residents.[16] Delhi claimed that it was releasing the thousands of monkeys it captured every year only in Asola.[17] However, the forest officials in Uttarakhand that I spoke with claimed that trucks full of monkeys were still being brought there from the National Capital Region (NCR).[18] In 2011, one official, who asked to remain anonymous, said, "the city [Delhi] has between 30,000 and 40,000 monkeys. Only half that number, probably less, are in Asola. Where do you think the rest are going? Delhi doesn't have enough forest to accommodate all these monkeys. Asola itself is like a desert."[19] When I asked if he thought that the Delhi forest department was involved in sending monkeys to other states, he suggested that monkey catchers were told by the forest department to release the monkeys in a region where they would no longer be Delhi's headache. Another official in Uttarakhand suggested that while the Delhi forest department might have initially sent captured monkeys to Asola, escapees from the sanctuary were being rounded up and sent to other states. A fellow forest officer in Uttar Pradesh, he told me, had recently informed him that more than a hundred monkeys had arrived in the state from Delhi under the cover of night. But, of course, Delhi was not the only culprit when it came to translocating difficult simians. Many state officials in Uttarakhand claimed that the monkeys being dumped in the mountains were also coming from Uttar Pradesh in large numbers. At the end of 2015, it was reported that the Principal Chief Conservator of Forests in Uttarakhand had written to the Director General of Police asking him to step up surveillance of state highways and take strict action against anyone caught transporting monkeys from other states. At the same time, the Forest Minister of the state was quoted as saying that the rise in Uttarakhand's monkey population and the resultant surge in complaints from villagers could be explained by the fact that "vehicles from neighboring states have been secretly releasing groups of monkeys into our forests in the middle of the night."[20] Several forest officials expressed serious anxiety that Uttarakhand's human population was at risk of contracting tuberculosis from monkeys they suspected were carrying the disease. The fact that monkeys captured in Delhi had been reported as carrying strains of tuberculosis strengthened the case made by these officials that the sudden spike in Uttarakhand's macaque population over the last decade was due to monkeys being translocated to the state from Delhi and Uttar Pradesh.[21]

Villagers who had long insisted that the monkeys who assailed them were brought to villages by the truckload from elsewhere were vindicated by this public confirmation from the state. "We have been saying this for years," one man said to me a few days after the minister's comments were reported in the press. "They just didn't have the courage to say it earlier. Politics." The sore point for villagers was that the monkeys were coming not just from other states but from cities such as Almora, Nainital, and Haldwani that were located within the state (Govindrajan 2015c). I was told by a senior forest official in Nainital that monkey catchers had been brought in from Mathura and Vrindavan to capture the monkeys—both rhesus macaques and langurs—who had attacked and bitten so many people that residents of the city lived in a constant state of panic. Early attempts to kill the offending animals had met with fierce opposition from Hindu nationalist groups and animal-rights organizations, forcing the forest department to opt for capture and translocation as a solution to the problem. The forest department insisted that the monkeys captured from Nainital had been released into a patch of reserved forest, but as villagers soon found out, rhesus monkeys preferred living around and with humans to staying in the forest. A similar situation arose in the villages around Binsar Wildlife Sanctuary when the municipal authorities in Almora captured its monkeys and released them into the sanctuary only to find that the monkeys promptly moved to and settled down in areas inhabited by humans.[22]

In his research on Balinese temple macaques, the anthropologist Agustín Fuentes (2010, 607) argues that the "interface" between temple macaques and the humans with whom they interact on a daily basis "constructs mutual ecologies that structure their relationships. In these zones of contact there is an entanglement of economies, bodies and daily practices that affect the size of macaque populations, their group compositions, and behavior." The monkeys who were translocated from cities, whether in the plains or in the mountains, to mountain villages were similarly accustomed to cohabiting with humans, what Fuentes, borrowing from Donna Haraway, calls "naturalcultural contact zones." When removed from such complex niches, urban macaques seek out the closest humans and create new but familiar zones of contact that resemble the ecology to which they are adapted. In effect, these monkeys are a "weed" species, a concept offered by a group of primatologists who suggest that macaques who successfully adapt to human presence are like weeds who thrive alongside humans (Richard, Goldstein, and Dewar 1989).

For the humans who had to live alongside these monkeys, however, there was not much comfort in knowing that they had been drafted as coparticipants in the construction of a new niche. While having to contend with monkeys from Delhi and Uttar Pradesh was, for many, the ultimate indignity, dealing with monkeys from Haldwani and Nainital was not that much better. One woman put the matter thus: "yeh pahar ke baanar honge lekin asli pahari nahi hain" (they might be monkeys from the mountains, but they are not truly monkeys of the mountains). When I asked her what she meant, she said that they were too urban (*shehri*) to be properly *pahari*. Their habits were akin to city folk, and they did not know how to behave in villages. For her, *pahariness* was linked to a particular kind of rural civility that was shared by humans and animals alike; the multiplicity of different *pahari* bodies were united by a "uniculturalism" (Viveiros de Castro 2004). Urban monkeys were unaccustomed to this form of civility even if they came from within the state, and this was what made them such difficult and unwelcome neighbors. They were, as one man put it, too *junglee* (wild) to live alongside *sabhya* (civilized) people and animals. Indeed, the idea was ubiquitous that these outsider monkeys, even if they came from as close as Nainital, were entirely removed from the affective circuits that characterized *pahari* interspecies sociality. The interlopers were universally decried as "aggressive" and "dangerous," animals who displayed no fear.

It is not entirely surprising that the people who encountered these monkeys in the mountains were so struck by their hostility. To understand why urban monkeys and translocated monkeys in particular are sometimes more aggressive than their rural cousins, we must turn to the particular history of the rhesus macaque in twentieth-century India. From the 1920s onward, juvenile rhesus macaques in India were in high demand for biomedical research in facilities across the United States.[23] Iqbal Malik, who has worked closely with monkeys in India for over thirty years, argues that the "haphazard trapping of individuals" from closely knit troops in this period produced what she calls a "chaotic fissioning" where monkeys formed smaller units for safety and dispersed over larger areas in search of food, eventually settling in urban areas.[24] This history explains why so many rhesus macaques in India live in and around human settlements. Malik suggests that the breakup of the troop during capture changed not only the "natural" habitat for these monkeys, but also affected them psychologically. She notes that "suckling youngsters separated from their mothers became depressed, while the mothers got more aggressive."[25] As Daniel Allen Solomon (2013, 98) observes, it is feasible that the "widespread fracturing of the troop structures would have long-lasting social effects in those rhesus populations who were most afflicted by trapping, effects which might resonate through the generations as learned behaviors."

Contemporary trapping practices continue to affect such disruptions. In 2011, I spoke with a monkey catcher from Uttar Pradesh who said that it was extremely difficult to trap whole troops together because the adult males and some females often ran away, leaving behind only juveniles and some adult females. On many trips, he said, he would catch only juveniles and subadults who would then be released in distant forests. In Delhi, a monkey catcher that the anthropologist Ajay Gandhi (2012) spent time with would sometimes capture babies who were separated from their mothers in the process. When individuals are separated from the troop and forced to create a new social structure in the unfamiliar area to which they are relocated, they must respond to new emotional and logistical challenges, a process of learning that often makes them rather bellicose. The primatologist John Gluck (2016) notes that the effect of social deprivation on captive young rhesus monkeys who were separated from their families was that they could not functional normally, and suffered "unimaginable anxiety, fear and depression." While the monkeys captured and translocated to the mountain villages of Kumaon were not hermetically separated from all others of their kind, the breakup of their kinship bonds must have had distressing effects in their social and emotional lives.

The *pahari* villagers who had to deal with the fallout of this complicated history, however, were largely unsympathetic to the plight of the monkey individuals whose social lives have been torn asunder. While they certainly believed that the histories and experiences of these monkeys were crucial to who and what they were as individuals, it was not the particularities of capture that they blamed for the aggression and wildness (*jangleepan*) displayed by outsider monkeys but the unsavory humans with whom they had kept company for so long before their exile from the city. For them, the monkeys' indecorous conduct revealed that their rightful place was in the Bhabhar alongside *junglee* (wild) plains people who were similarly misbehaved; theirs was a fellowship of *gundas* (goons), as one man put it. This "connective language" is revealing of how critical exclusionary forms of difference were to this politics of insider/outsider (Weaver 2015).

For many, then, what made these monkeys so unwelcome was their striking difference from *pahari* monkeys and people. There was a pervasive belief among villagers that it was no coincidence that *paharis* were the ones called on to bear the burden of hosting these monkey strangers with their frightening belligerence. While some said that this was yet another example of how *paharis* had to suffer so that plains people could prosper, other people argued that the sudden appearance of the outsider monkeys was not simply a matter of the plains displacing their problem to the hills. These were not ordinary animals, they said, and their presence in the mountains foreshadowed something darker to come.

The Urban Monkey as Criminal Human

"The monkey terror only keeps growing." I was talking to Pushkar Bisht of the Gwal Sena over the telephone a few days after learning that the organization was planning to involve itself in the matter of the monkeys. When I asked him about the monkeys, he assured me that the monkeys were not from the mountains but were part of an elaborate conspiracy (*saazish*). "They are agents of the land mafia," he said. When I sounded dubious, he explained that the monkeys were captured from cities as far as Delhi and Mathura and released into mountain villages by members of the land mafia. This was not a small matter, Bisht cautioned me. The people who were buying land from *paharis* were from Delhi, Kolkata, and Mumbai, but the people they sold the land to in turn were local representatives of "internationals" from China and Pakistan. Even the D-Company, he told me, had its local agents working to acquire land in the mountains through extortion and deceit; they too were part of the monkey conspiracy.[26] In some cases, Bisht went on, the land mafia paid monkey catchers employed by the forest department in Delhi, Uttar Pradesh, and even Uttarakhand to release the monkeys they captured in urban areas near mountain villagers. There were photographs, Bisht claimed, of people unloading monkeys from SUVs. How did they get there? Well, who else but the land mafia would bring monkeys in their cars. Their strategy, he told me angrily, was working. Every day, people were abandoning cultivation because of the depredations of the monkeys. Many had sold their land and were leaving the mountains. *Paharis* were leaving in droves and meanwhile the *pahar* was being despoiled by outsiders. For him, the monkeys' intentions were unimportant; what mattered was that their actions allowed a larger chain of violent events to unfold.[27]

When I talked to Bisht again a few years later, he was still absorbed with his task of battling the monkey menace. He was upset that animal-rights organizations were blocking people's right to defend themselves against these invaders and repeated his views on the importance of punishing animal criminals (Govindrajan 2015c):

> These animal-rights groups like the PFA insist that we not lay a hand on these monkeys. I think that's wrong. They don't know the terror *pahari* villagers live in. At least we should have the right to wound these monkeys. Criminals are beaten by the police. If these monkeys steal and engage in hooliganism, then we should have the right to punish them too. They are criminals [*apradhi*], not animals; the people who commit the gravest of crimes [*sabse sangeen jurm*] teach them how to thieve. What is the difference between them and humans then? Go to the plains and see for yourself how they steal from people. In the mountains these outsider monkeys are completely unafraid because we sit here with our hands tied while they engage in their criminal activities.

I was both fascinated and disturbed by Bisht's strident views on the criminality of emigrant monkeys. Theirs was not the lovable and ultimately forgivable act of theft committed by *pahari* monkeys. On the contrary, the outsider monkey was a hardened criminal whose illicit acts mirrored those committed by human malefactors. Interestingly, Bisht did not believe that this criminality was the outcome of an animal nature, an association that has a long and complicated history in both colonial and vernacular traditions. As Anand Pandian (2009, 103) points out, Enlightenment thought, in particular, depicted the project of "perfecting human nature" as one that involved overcoming one's bestial nature. In the colony, the effects of this reasoning meant that colonial subjects who were judged less than human "either had to submit themselves to ambitious projects of training, discipline and domestication or to endure the implacable violence of an exclusionary humanism." Pandian (2009, 103–5) notes that contemporary understandings of moral underdevelopment as the outcome of an uncontrolled animal nature in the Cumbum Valley, where he did fieldwork, are "indebted [both] to the legacies of colonial criminology and south Indian moral tradition," which also emphasizes restraint as a marker of human nature. However, in Pushkar Bisht's moral imaginary, the problem was not that criminal humans and simians lacked the particularly human "virtue of restraint." On the contrary, the criminality of the outsider monkeys was explained by the fact that they were no longer animal and indeed were so tainted by the acquisition of human tendencies toward delinquency that there was no difference between them and the human criminal. As criminals, Bisht was arguing, they had to suffer the consequences of their thievish and loutish actions.

For some, the concern was *where* and from *whom* these outsider monkeys had picked up these vices. The real trouble, one old man observed, was that these monkeys had learned their immoral ways from plains people (*bhabhar-walleh*) who excelled in "thuggery." "Our monkeys are like us," he said, "rank simpletons (*kuch zyada hi seedhe*)."

> But these outsider monkeys have been raised by the most cunning people in the *bhabhar*. My brother had his glasses stolen by a monkey in Mathura. He was wearing them and suddenly a monkey came out of nowhere and just snatched them off his face. He had to pay a beggar boy who was watching the scene twenty rupees to get his glasses back from the monkey. Ha! Beggar indeed. They must have been working together. Such criminal tendencies at a young age. Everyone in the *bhabhar* is like that. Now you tell me, if an animal is so good at this work, is he an animal any longer? Finding an animal in the plains has become hard nowadays, they have all become human (*insaan*). All the vices (*buraiyaan*) of humans are in them. That's why this matter is so dangerous. People say "they are only monkeys who are doing the work of monkeys." But no, these monkeys who have come from elsewhere are actually doing the work of humans—they are thieves, dacoits, goons.[28]

Here too, as in Bisht's account, monkeys from the lowlands are considered unworthy of forgiveness for acts that in the case of the *pahari* monkey might be considered mischief. Instead, having lost their animal innocence as a result of too much time spent in the company of the deeply mistrusted *bhabharwallahs*, they are thought to be possessed of a particularly human criminality. What made the matter of the monkeys so desperate, then, is the fact that many people believed they were not dealing with ordinary monkeys but with monkeys who had taken on the very worst qualities of humans.

What is so striking about the old man's insistence that these thieving monkeys have become human (*insaan ban gaye hain*) is his imagining of human and monkey bodies as sites of exchange that are porous, open to being affected by the other. This is, as we have seen in chapters 2 and 3, a conception shared by villagers who tend goats and cows and who recognize that their subjectivity is formed in relation to the animals who reside in their "near sphere." Becoming, in other words, was widely recognized as a transcorporeal process of exchange. What this meant for the old man was that while the boy in Mathura might have initially taught the monkey how to thieve, the monkey was soon a true accomplice, having imbibed and embodied the boy's criminality. Who knows what the monkey might have taught the boy in turn, the old man re- a little later in our conversation. It is precisely because their criminality e unlearned, having been so thoroughly impressed on their bodies e of encounter, that the outsider monkeys were considered so dangerous. There was no doubt, in some people's minds that these monkeys had to be punished as humans would have been for similar acts of destruction, for they were motivated by the same criminal impulses. *If an animal is so good at this work, is he an animal any longer?*

In practice, however, the answer to that question remained complicated. On one occasion, I was waiting for a bus when conversation among the others at the bus stop turned to the monkeys who were recent arrivals to the next village. When I mentioned that farmers in the adjoining mountain state of Himachal Pradesh had recently held a mass protest demanding the right to shoot monkeys that raided their crops, the man sitting next to me, a former soldier in the Indian Army, was enthusiastic at the prospect of monkeys being "taught a lesson." "People in Himachal," he declared, "know how to fight for their rights. We should also do this—get out our guns and kill." A younger man, whose hair was slick with styling cream, protested: "Bandaron ko maarne se paap lagta hai (killing monkeys is a sin)," he warned.[29]

The army man was unimpressed. "These monkeys are not Hanuman," he scoffed, "they're demons [*rakshasas*]. I've seen many *gundas* in my lifetime, but they are the most dangerous." I was reminded in that moment of how fragile and partial Hanuman's protection can be for the simians who are often worshipped in his name. At that point, another voice intruded into the conversation. It came from a man sitting on the road barrier. "Yes, they are *gundas,*" he said, "but bad things will happen if you try to kill them." He told us the story of a boy who had tried to trap a monkey in his house and hit it with a stick. Even though the monkey escaped largely unharmed, the boy was punished for his attempt a few days later when a group of monkeys who had been lying in wait for him by the road mauled him. "They're not ordinary monkeys," he conceded, "but maybe that's why killing them is more dangerous." There was silence as we contemplated his words. The army man was still unconvinced, but the others seemed quite taken with the argument that these monkeys needed to be treated carefully, not just because of the questionable moral and religious ethics of killing monkeys but also because there was something peculiar and unfamiliar about these particular monkeys. Just as when dealing with unpredictable and potentially dangerous plains people, one had to be on one's guard with these outsider monkeys who were becoming more and more like their slippery human cousins.

*

"By the time these people decide what to do, we will be finished." My friend Kusum's tone was dark as she responded to the news that the Uttarakhand government was once again contemplating a sterilization program to control

monkey populations in the state. The first time around, the government's an-
nouncement that it would open two centers, one in Kumaon and the other
in Garhwal, for the training and rehabilitation of troublesome monkey populations was met with opposition from Maneka Gandhi and the PFA on the
grounds that they would entail unnecessary cruelty to the animals. Gandhi
argued that the centers were simply fronts for ill-conceived sterilization cam-
paigns that would end in failure as had been the case in the state of Himachal Pradesh, where botched surgeries led to hundreds of monkey deaths. In
an editorial on the growing conflict between humans and animals, Gandhi
explained why she was opposed to programs of monkey sterilization, citing
not just their cruelty but also the damaging effects on monkey sociality and
kinship:

> Monkeys live, like humans, in large extended families. If you trap a single
> monkey, then two things happen: the family rushes away from their own terri-
> tory/area. They get very angry with humans. So they enter areas that they are
> unfamiliar with, because now they are too scared to go back to an area where
> their relative was trapped. Now they enter people's homes and fields in search
> of new larders and in the process hurt humans and themselves and destroy
> crops. The single monkey that has been captured usually breaks a limb during
> this hideous capturing process (Himachal pays Rs.500 to anyone who captures
> a monkey) and is sterilized while in terrible pain. He/she usually dies (the
> mortality of Himachal's monkeys by the so called vets is unbelievably high).
> If he/she survives, it is left at the place where it was picked up. But by then
> the family has disappeared and the troupe replacing them will not accept this
> poor animal. So, to survive, he enters human homes and becomes predatory.[30]

In the face of such opposition, the government changed tack and encouraged the forest department to plant fruit trees in forests so as to attract monkeys away from villages, a move that was applauded by the PFA. Unfortunately,
that plan, too, was a dismal failure, and what was popularly known as the
monkey menace continued unabated. 2015 was round two of the sterilization
solution, and no one was particularly optimistic that it would work this time
either. The government's plan to use contraception was greeted with general
mockery. "They should start distributing Mala D [birth control pills] to village women and *pahari* monkeys instead," Lata, a newly married woman said.
"The only solution to this problem is that all *paharis* should die so that these
outsiders [who are buying land] can divide the village between themselves."

Lata was being facetious, but it was true that villagers were fast despairing
of this barrage of solutions. Things were not helped by the fact that there

was an intense moral panic around the growing rates of migration from villages across Garhwal and Kumaon (Govindrajan 2015c). While migration has been a historically important livelihood strategy in the mountains, the scale of contemporary migration, as I observed in the introduction, was held up by observers as a cause for alarm. Some villages had been entirely abandoned as their residents moved to cities in the lowlands to eke out a living. These desolate "ghost villages" had become the emblem of this crisis in the local and national media, their state of ruin providing a glimpse of what was in store for other villages across the mountains in the future.[31] A chronic lack of employment and opportunity was usually touted by villagers as the reason for their departure from the mountains. Agriculture, which had always been a difficult prospect in this unrelenting terrain, had been further hit by the shortage of agrarian labor that resulted from growing rates of migration.[32]

As people abandoned cultivation and moved to cities, their fields returned to nature, providing a haven for wild animals who moved into these new microhabitats located in and around villages (Govindrajan 2015b). Even those who remained in the village stopped farming in the vicinity of these abandoned and overgrown fields for fear of encountering dangerous wild animals. This domino effect contributed to an agrarian crisis across Kumaon with some villagers ceasing all cultivation even if they continued to live in the village. The idea that cultivation for the market was not worth the effort was ubiquitous during my fieldwork.

This difficult situation was exacerbated by the fact that youth in the mountains, who viewed agricultural labor as debasing, were increasingly disinclined to spend their life toiling in fields even if it meant being unemployed until something better came along. The work provided by the Mahatma Gandhi National Rural Employment Guarantee Act (MGNREGA) was also regarded by many as too demeaning for the educated and too low paying for those who were willing to work given that the daily wage rates on private construction sites were much higher than those offered by the state under the act. For these youths, then, migration was as much a lifestyle choice as it was a livelihood strategy. Cities were seen as spaces of modernity, while life in the village represented backwardness and failure. I heard over and over again that opportunities for development could only be found in the lowland cities of Uttarakhand, not in its mountains. This was why, people said, growing numbers of youths were leaving the mountains every year. As this process unfolded, organizations such as the Gwal Sena explicitly linked the lack of *pahari* control over resources like *jal, jangal,* and *zameen* with the loss of *pahari* identity. It was only when *paharis* would fully control the mountains and its resources, they said, that the promise of Uttarakhand would be realized. Others, too, drew similar connections between maintaining cultural and political integrity and controlling access to land and resources.

The matter of land and water was especially complicated because outsiders were moving to the mountains in search of property as fast as *pahari* youth left them. Private real-estate accumulation expanded dramatically in India since the liberalization of India's economy in 1991.[33] In Uttarakhand, too, the years since 2000 saw a sharp rise in the sale of agrarian land to rich Indians from cities in the lowlands. Even when the real-estate market slowed down in 2012, there were still a good number of transactions in land taking place in the villages that I visited. Fueled by a romantic bourgeois discourse that represented the Himalayas as a pristine locale far removed the hustle and bustle of city life, people were flocking to Uttarakhand to buy land for second homes, or to flip when prices rose.[34] Although farmers participated willingly in this land market, the scale and pace at which land was changing hands was enough to raise the eyebrows of even those who had been among the first to sell their land to buyers from the lowlands. In Pokhri, villagers would joke that a cluster of houses at the top of the village belonging to people from Delhi, Gurgaon, Mumbai, Lucknow, and Nainital would confuse visitors into thinking that they had arrived in a big city instead of Kumaon. Many would say *hamara gaon ab Gurgaon ban gaya hai* (our village has become Gurgaon now). It was against this background that the sale of ten acres of prime land to a real-estate developer who planned to build 150 studio apartments on the property caused tremendous consternation.

Some were afraid of another epidemic like the cases of typhoid that rapidly spread through the village during one monsoon season when the feces from a temporary settlement where laborers on a housing project had defecated in the open was swept into drinking water sources. Others were worried about the effect the new arrivals would have on the delicate social ecology of the village. "It's difficult enough to share water with all the people who have already moved here," an older villager said to me, "but this will destroy the village entirely. We will have no space, no water. Our air will no longer be clean and pure. Where will people graze their cows and goats? There is destruction written in our future." Most villagers shared his views on the impossibility of so many people living together in a space without the destruction of the commons. During an acute water crisis one year, the fact that some outsiders—hoteliers and homeowners—survived the drought by purchasing multiple tanks of water drawn from a source held in common by the village was confirmation for many that in a battle between villagers and outsiders,

FIGURE 21. The site of a proposed apartment complex in Pokhri.

the wealth of the latter would ensure their survival. "In ten years," Deepa, a young Dalit woman, said to me, "not many *paharis* will be left in this village."

> It's too expensive here now. We can't afford to make repairs to our house because these outsiders have driven the price of materials and labor up. Our kitchen is leaking, but we don't have the money to redo the roof. There's no water in any of our *naulas*; they have all dried up. Ok, there hasn't been much rain this year, but so many people are still taking water to construct their *kothis* (bungalows).[35] We can't stop them because they are rich and know all the important state officials. There is no water left for us. We have to walk so far to bring water from the hand pump. My husband has already been talking about moving to Haldwani. Life in this village is becoming harder for us by the day. Only outsiders can pay the price to live here now.

In other villages, too, battles over land and water were escalating. In Danka, Murli told me how his younger brother had betrayed him at the instigation of an outsider who wanted to buy land in the village. The brothers had swapped land two decades or so ago, before the prices of land in the mountains had risen so steeply; Murli had taken the family land in the mountains and, in return, offered his younger brother a plot of land in Haldwani. Even though the plot in the mountains was much larger than the plot in Haldwani, they were of an equivalent price at the time. The brothers had not bothered to change legal title; after all, as Murli reasoned, they were family. But recently, Murli's brother had announced that he was selling the plot in the mountains to a real-estate developer who wanted to build a hotel on it. He told Murli that he could have the plot in Haldwani back; the brother himself was moving to another, bigger plot in Haldwani that he had purchased with the advance given to him by the would-be hotelier. Murli was furious; he had spent thousands of rupees on making improvements to the land in the mountains, while his brother had allowed the land in Haldwani to fall into disrepair. But there was nothing he could do, at least not legally. He decided to take the matter to Golu *dyavta*'s court, a solution I had seen others resort to in land disputes (Govindrajan 2013). Murli blamed the situation on the outsider who had led his brother astray. "These outsiders will do anything for land in the mountains," he said. "Who knows what he said to my brother to convince him. He must have given him a lot of money. The saddest thing for me is that we no longer have any land in the mountains. Our relationship with the mountains has been snapped."

Even though villagers did appreciate the presence of some outsiders, especially those who made the effort to know them as individuals and establish relationships of trust and reciprocity with them, the majority of new arrivals were often blamed for fraying the bonds of relationality that were crucial to the reproduction of family and community. Some outsiders, especially real-estate developers who were often involved in shadowy deals and hired local toughs to "'assist" them in case things went wrong, were singled out as duplicitous and dangerous. Mohan, a man I met at a wedding in Almora district, told me how he had been cheated out of a large sum of money by a buyer from the city of Moradabad. The developer had paid him an advance and promised that the rest of the payment would soon follow. In the meantime, he suggested that it was expedient to go through with the legal process of changing title lest his partners get cold feet. When Mohan recounted this part of the story to me, he grimaced at the memory of what had followed.

> I was so stupid. I thought that such a big man would never cheat me. He was so nice; he brought books for my children. But then after I registered the land in his name, the rest of the money never came. I used to call him a hundred times a day. He never picked up the phone. When I sent him an SMS threatening to call the police, the *gundas* came. They broke my hand and said worse things would happen if I pursued the case. In the end I gave up. There's no point in going to the police; he can buy them easily. I thought I would buy some land elsewhere with the money. But now all I can afford is a small plot, not big enough for a farm. My wife says we should take a room on rent in Almora,

and that I should find a job there. But this is my ancestral village. I don't want
to leave it.

Such stories of betrayal abound. Things, people said, were not helped by
the fact that village youths were seduced by the outsiders' evident wealth and
power and imagined similar futures of moral and material privilege for themselves. What the youths did not realize, older villagers often reflected, was
that money never came free; eventually, things would go wrong. There were
those who had gone into the land business—buying land from locals to flip to
outsiders at a higher rate—and overextended themselves; they were stuck with
large plots of unsold land, especially after a national slowdown in the real-
estate business from the end of 2012. There were others who had been cheated
and had no recourse. Some who had managed to sell their land without any
complications found that the money from the sale did not last as long as they
had expected. They were forced to borrow money from their relatives or had
to return to the mountains to beg for a share in the extended family's property.
These numerous examples led many *paharis* to conclude that their situation
would not have been so bad if they hadn't been tempted by the prices outsiders
were willing to pay for their land. After all, land was something to fall back on
in difficult times; without land, one was forced to beg or steal when money, a
fickle lover, ran out.

The real-estate boom in Uttarakhand had thus produced a world in which
the opportunity to turn a huge profit on land sales was tempered by the fact
that one could be cheated by those who were more skilled at navigating this
complicated world than the ordinary *pahari* villager. There were many who
feared that the mountains would soon be emptied of *paharis* like Mohan who
were deemed too innocent to navigate the web of deceit spun by outsiders who
wanted their land at any cost. One of the most nefarious strategies employed
by outsiders, people like Pushkar Bisht of the Gwal Sena believed, was the
release of aggressive and fearless monkeys into villages. The monkeys soon
established a reign of terror that forced people to stop cultivation altogether
and eventually sell their land. Even those who doubted that the land mafia was
responsible for bringing the monkeys to villages were struck by the fact that
so much *pahari* exodus from the mountains had been fueled by the influx of
both human and simian newcomers. Perhaps it was the coincidence of their
arrival or the belief that it was no coincidence at all, but outsider monkeys and
outsider humans seemed to share much in common, especially their ability to
make life in the mountains difficult for *paharis*.

At War with Strangers

It was the time of year when the pear trees were laden with fruit, their boughs hanging low to the ground from the weight. After coming down in relentless torrents for three days, the rain was finally spent. Now, only the sound of monkeys screaming and chattering as they fought over fruit penetrated the steamy cloud that had settled over us, a signal to the human residents of the village that it was time to come out from their houses and protect their orchards. An eighty-four-year-old man, whom I knew simply as Bubu (grandfather), emerged reluctantly into the open and walked up the short path to the highway on his short, bowed legs. There they were, sitting squarely in the middle of the narrow ribbon of highway: a group of young rhesus males, females and matriarchs, and lots of little babies, all feasting on fruit that a few of them had shaken to the ground. Bubu let out a pained groan at the sight. He picked up a few stones and flung them at the macaques. They ducked the stones with easy grace and remained where they were, eating a little bit of each pear before flinging it away and starting on another.

I watched as one of the monkeys, a juvenile whose dark-red fur was patchy from an attack of scabies, ignored the fruit lying in front of him, jumped up on a low stone wall on the side of the road, and shook more fruit to the ground from the low branches of a beautiful old tree. On most days, he could be found swinging on the branches of that same tree. When the monkeys had first shown up in that area, some of the older females used to chase this juvenile—whom I called Lalu for his color—every time he tried to get too close to them and their babies, but now one of the smaller groups had mostly come to accept him and would let him play with their babies. One no longer heard the plaintive, distressed calls he would utter as he sat alone on top of the tall pine tree near the orchard; instead he would spend all day frolicking with some of the other juveniles.

Meanwhile, Bubu was incandescent with anger. The abuses he hurled at the monkeys now had more force than his feeble throws—*randichyol* (sons of whores), *haraimi* (bastards), *kukur* (dogs). The macaques scattered as Bubu walked among them, shaking his fist and swearing revenge, but regrouped as soon as he had passed. So it went for the next half an hour, Bubu dispersing the monkeys with wild swings of his cane only to turn around and see that they were back in their places—a few perched precariously on the loose slate roof of his house; two mothers and their babies sitting comfortably on packed crates of fruit cooing to each other between bites of juicy fruit; some concealed amid the leaves in the vivid green forest of trees. I was impressed by how unperturbed the monkeys were by his presence; a few even threatened him

FIGURE 22. The group of outsider monkeys whose raids on Bubu's orchards were particularly damaging.

with an open mouth when he walked by. "Saale bilkul darte nahin, bahar se aaye hain naa" (these bastards have no fear, they have come from outside) he shouted to me, when he noticed me watching the scene. "Fling a stone at that one in the trees." I obeyed somewhat reluctantly, fearful that the macaques would give chase and bite me, as had happened to someone I knew. Thankfully, the pebble I flicked went wide of the tree, and the rhesus concentrated on eating with nary a glance in my direction.

I walked over to stand with Bubu, who had now given up on chasing the macaques and was instead looking at them despondently as they demolished his fruit. "They have eaten at least thirty crates worth of fruit," he said, defeat writ large on his fragile body. He recalled the time when the monkeys first appeared a few years ago. It was the night that his neighbor had heard trucks drawing to a stop in the little patch of pine forest that separated their cluster of houses and shops from the next village. The neighbor thought that passing truckers had brought prostitutes to have sex with in the forest and decided not to check on the sounds. Around midmorning, about forty rhesus macaques descended on the village. They were not the forest macaques that people knew well; those monkeys, Bubu told me, lived mostly in the forest, and still occupied their old niche. The new macaques, he said, were redder and had shorter hair than mountain monkeys. And they were completely unafraid. Forest monkeys, if chased, would turn tail and run away. They are shy (*sharmila*) monkeys, he said, just like *pahari* people.[36] But these city monkeys were aggressive and pursued *you* if you had the temerity to threaten them. Even an air gun fired in their general direction succeeded only in scattering them a few paces. Unlike forest macaques, who were occasional visitors to the village, these monkeys were unwanted residents who had not left the village once in the three or four years they had been there. In primatological terms, these monkeys were fully synanthropic (Gumert, Fuentes, and Jones-Engel 2011). Since their arrival, Bubu reflected, things had not been the same. One of his neighbors had recently sold a large part of his farm to a retired couple from Delhi. The neighbor had told Bubu that he never wanted to spend another day on a futile chase through the orchard; there was no point, the monkeys would still eat everything. These were not the monkeys of old, Bubu said thoughtfully. They had managed to defeat everyone who tried to stand in their way.

I, too, had noticed several differences between these "newcomers" and the rhesus macaques I encountered deep in the forest at some distance from the village. While the forest macaques would all sleep in the same tree, the macaques in Bubu's village would scatter and climb into different trees when it was time to sleep; it was only after a few years of being in the village that they finally started sleeping in the same tree. If they spotted a human walking through the woods, the forest monkeys would retreat deeper into the dark embrace of the trees; one could hear them crashing through leaves and branches as they made their way to safety. The macaques who lived near Bubu, on the other hand, had no compunction about being around humans; they preferred to sit on the road and would barely move even when a car would drive by within touching distance of them. The forest monkeys spent more of their day grooming each other compared with the village macaques and were also less aggressive than the macaques who had settled near Bubu's village. On the occasions that I was able to watch them at length, the monkeys in the forest fed calmly next to one another with only a few fights breaking out over who would feed where. The newcomers to the village, on the other hand, would frequently threaten, chase, and bite each other. The display of threatening behavior followed by chases and attacks were an almost constant feature of their days, although they did grow to be more cohesive over the five years that I knew them. Fights were at their most bitter when monkeys competed over food stolen from people's houses or the occasional offering from tourists who were passing through.

Primatologists who have spent time observing urban macaques and langurs argue that increased aggression and competitiveness are features of macaque social life in areas densely inhabited by humans since reliance on anthropogenic sources of food is high in these contexts.[37] However, it is also possible, as Ciani (1986, 438) suggests from his observations of a single group

FIGURE 23. Villagers claimed that the behavior of the outsider monkeys was a dead giveaway as to their provenance.

of rhesus macaques who moved between forest and town in Simla, that these "outsider monkeys" were actually forest macaques "who changed their behavior to exploit resources from . . . two very different habitats."

All the villagers I spoke with, however, were certain that these animals had come from elsewhere, a claim that is believable not only because of the murky politics and history of monkey translocation in north India but also because these were people who *knew* the animals with whom they lived intimately. In many cases they had spent several decades in the same place and had taken the time to learn about the distinctive characteristics and tendencies of the animals whose paths crisscrossed theirs. Their confidence that these animals were outsiders was grounded in situated observation of them as individuals whose lives were entangled with their own, not on simple flights of fancy.

Indeed, if anyone expressed doubt that these monkeys had actually come from elsewhere, villagers were quick to point out that the bodies and behavior of the macaques constituted the most compelling clue to their origins. On one occasion, as I sipped a cup of tea in a shop located in Bubu's village, the monkeys descended on us. The tin roof rattled with a terrifying ferocity as they leaped onto it from the surrounding pine trees. Every few minutes, a monkey's inquisitive face would appear over the awning and assess what could be most easily stolen. The shopkeeper stood by his wares with a determined look on his face and a stout stick in his hands. "You had better watch out for your Maggi," he said to me, warningly. "They love Maggi most of all, these bastards. Eat it faster than the tourists from the plains." I couldn't help but laugh. "What else do they like to eat?" I asked him. "Their *culture* is different than ours," he responded. "They are used to eating noodles and chips. That's what people eat in cities, and that's what these monkeys have learned. Here, let me show you." With that, he put a packet of Lays chips (Tomato Tango flavor) on a bench just outside the shop. The moment he left his vigil at the storefront and retreated inside, a young monkey jumped onto the bench from the roof and grabbed the packet. She took it across the street and managed to puncture it with her teeth. I was envious of her dexterity as I recalled how many times I had struggled to open one of these packets. We watched as she licked the spice off some chips. Her eyes would close momentarily with pleasure each time her tongue found a grain of salt, but then dart around to make sure no one was coming to deprive her of her snack. I was delighted by the enjoyment writ large on her face. When she was done she ripped the packet open and licked the foil clean before throwing it aside and running off. "Even our children wouldn't be able to open the packet with such ease," the shopkeeper said half-admiringly, "and they eat this rubbish all the time. I'm telling you, the *culture* of these monkeys is different. A *pahari* monkey would never know how to open this packet. But these monkeys, they will not steal *chana*, they go straight for the cream biscuits and the chips. They will eat a roti if it is all they can get, but if they find some Maggi, then its *wah wah* for them. Their tastes are just like these tourists from the plains. Gluttonous [*khadu*]"

On another occasion, a woman who now cultivated only ginger because the monkeys had eaten everything else she had planted for two years in a row told me about coming home to find the monkeys in her kitchen.[38] They were sitting around the fire that her teenage daughter had built in the kitchen hearth and eating the *subzi* and *roti* that she had prepared. The daughter was hiding in one of the other rooms, waiting for someone to come and scare away the monkeys. It turned out that they had surprised her in the kitchen and chased her out. She was shaking with fright having narrowly escaped being bitten. "The monkeys that would come to our house earlier would never do that. They would run in and steal something from the garden only if no one was looking. These monkeys have no fear. We are all so scared of them now that if we see them sitting in a group, we take another route." Another man told me that he had found a monkey sitting on the bed in front of his little red mirror that was propped up in the kitchen window frame. The monkey was combing her hair with his wife's comb and admiring her reflection much as his wife did. "These monkeys are so clever," he marveled. "They are even smarter than we are. If you send us to the city, we won't know what to do, but these monkeys will. They know everything. Just as cunning as people from the plains."

Some people thought it likely that outsider monkeys would soon teach *pahari* monkeys their depraved city ways.[39] Even though forest rhesus were often driven away by these more aggressive outsiders, social and sexual interactions between different troops was not out of the realm of possibility. Certainly opportunities would arise for *pahari* monkeys to watch and learn from the techniques employed by their savvier urban cousins. A few compared the looming ruin of *pahari* monkeys to the ways in which *pahari* youth had been spoiled by learning city ways through television but also through lived example in the form of their new neighbors from the city. One woman who had been particularly harassed by the monkeys told me that two groups of monkeys had now descended on her farm. "Some of them have short hair, some have longer hair," she said. "I think at least some of them are our *pahari* monkeys. Look at how easily they have been spoiled by these outsiders." Another woman who was sitting with her was particularly moved by this observation. She told us that her son had recently started working for a foreigner who had just rented a house in the village from a man who lived in Delhi. "He drinks so much, so much. And women come to his house when he is alone sometimes. I don't want my son to work there and learn these things. So many of our boys and girls have been ruined by their influence." They were both sunk in a gloom for a few minutes before the first woman spoke up again. "It's monkeys on one side and men on the other. These outsiders (*baharwalleh*) will leave us alone only when every single *pahari* has gone."

*

Allow me to return briefly to that rain-soaked afternoon with Bubu and the monkeys in 2014. As we stood by the side of the highway, surrounded by monkeys on all sides, I suddenly noticed a juvenile female creeping up to us and flinched. "*Hat*," shouted Bubu. The monkey cowered, but continued determinedly toward us. As she came closer, I looked instinctively at Bubu's cane, but it had not moved from his side. I was stunned when the monkey touched Bubu's leg with her delicate brown fingers. She was close enough now that the flecks in her hazel eyes jumped out at me. When she closed her eyes, I saw that her lids were almost blue. She looked up at Bubu with an expression that could not be described as anything but searching. I could see that she was young, her face only a little more mature than a baby's. She was beautiful. I wanted to reach out and touch her stubby fur, but my hands were pinned to my sides from fear and anticipation. This was the first time I had been in touching distance of any individual from this group. I was relying on Bubu to scare her off, but he was uncharacteristically quiet. Finally, with an air of obvious embarrassment, he picked up a fallen pear and handed it to the monkey. "Bubu?" My tone was a mix of accusation, amusement, and incredulity. The knowledge that Bubu, who spent so much of his day wishing fire and brimstone down on these monkeys had actually fed one of them was hard to digest. "I never feed them otherwise," he said, defensively. "But this one doesn't seem to have a mother. She usually sits alone, and all the others bite her. She started following me from the first day itself. I tried to hit her a few times, but she still wouldn't leave me alone. I felt pity (*daya*) for her. She might have come with the rest of them but she's not like them. She doesn't steal like them. That's why I give her something every now and then."

I remain moved by the affective force of that moment even today. What touched me, in particular, was that brief glimpse of tenderness in a relationship that was otherwise marked only by resentment and desperation. For me, that brief interaction between Bubu and the juvenile monkey, one of several such encounters, provided a vivid glimpse of how it is through these embodied, situated encounters that different creatures can pull one another into their orbits and make mutual claims that relate them to one another. I was dubious of Bubu's assertion that his affection for this monkey, unlike the others, was because she did not steal. But I do believe that something in her gaze—some expectation of recognition and reciprocity, perhaps—held his and allowed them to foster a connection that lingered despite Bubu's generally malevolent feelings about the outsiders; perhaps, her face held a mirror up to Bubu (Ohnuki-Tierney 1987). As I recall her fingers, so much like my own, the black nails standing out against the skin of Bubu's leg, I am moved by the "room" that such situated and embodied relationships between distinct individuals can make "for life . . . for possibility, for chance" (Ahmed 2010, 20). Love, as Donna Haraway so beautifully puts it, is risky precisely because it "undoes and re-does you" (Haraway in Gane 2006, 144). In the next chapter, I offer another instance of how the particular histories of certain creatures can destabilize conceptions of the world and offer possibilities for new forms of world making. Let us now turn now to the story of the pig who went wild.

Pig Gone Wild
Colonialism, Conservation, and the Otherwild

The runaway sow's story came to me unexpectedly and during an unfortunate event. I was helping Komal Arya carry bales of dried hay, fragrant with errant strands of wild thyme, to the shed where we would spread it in front of her cows and goats who had been bellowing their hunger for the last hour. As I made my way carefully down the steep concrete path that connected houses at the bottom of the village to the main road, we ran into Neema Bisht, our elderly neighbor. You could read the agitation that gripped her wiry frame. "Shivo," she exclaimed, her voice full of dismay and excitement. When I asked her what was wrong, she started at the fact that we had not heard the news until then. "You know Hema Arya from Sirmaur? She was attacked by a wild boar this morning." She stretched her hands apart to show us the supposed size of the boar's tusks. "Couldn't even run before it attacked her."

A visceral shudder passed through my body at the thought. "Was she badly hurt?" I asked.

Neema's tone was somber as she shared what she knew. "The 108 [the emergency number used for dialing an ambulance, used by villagers to refer to the ambulance itself] came. She might even be dead . . . who knows? Someone told me that the boar tore her open with his tusks."

She was walking to the village tea shop on the edge of the road, she told us, to find out more. After all, her daughter was Hema's friend from school. Taking the hay from me, Komal asked me to accompany Neema and get more details.

By the time we reached the shop, it was crowded with people who had just arrived on the afternoon bus. The air was thick with details of the encounter. "Their only daughter-in-law; who will work in the fields now that she's in hospital?"; "tusks the size of an elephant"; "I heard she was still alive when the ambulance arrived at the hospital, god knows how."

The elderly owner of the shop, Jeevan Bhatt, cleared his throat meaningfully several times as he poured yet another cup of *chaha* from a saucepan congealed with grounds from the day's endless rounds of tea. "Listen," he said. Slowly the babble of voices in the cramped shop, a sheet of tin suspended over four poles of delicate bamboo, ceased. "One thing is clear," he continued. "These wild boar (*jungli soongar*) are getting more dangerous every day. There was a time when we never saw them, not even in the forest. Now they roam the village by day . . . it's our food they eat, our bodies they tear. It's a bad situation. And we can't touch a hair on these bastards' heads or else the forest *wallahs* [forest department officials] will show up and then . . ."

"Two or three of my fields are near the forest . . . I've stopped visiting them," Radha, our neighbor, said. A few hours earlier, she had told Komal and me that her husband was looking for a buyer for those fields, preferably the same family from Kolkata that had just bought a plot of land in her village. "The last time I went there three or four pigs were lying in the grass nearby. I ran home and haven't gone back. Why put my life in danger for some half-eaten potatoes?"

Jeevan Bhatt cursed as a splatter of hot oil from the pan in which he was frying samosas landed on his forearm. Wiping it down on the front of his already grease-splattered shirt, he responded "Why do you think I started this shop? Who has the money just to feed pigs? Potato seeds, potatoes, cauliflowers, cabbage, peas, *madua* [finger millet]. They eat everything. I can't believe there was a time when there used to be hardly any wild pigs here."

This was the second time he had mentioned that wild boar were new additions to this landscape. "But where did these boars come from then?" I asked, raising my voice to be heard above everyone else chiming in to recount the damage caused to their fields by wild boar.

"About thirty or thirty-five years have passed . . . or a little more. How many years, Mahesh?" he asked, turning to the wizened old man sitting next to him and rolling his *beedi*. "I don't remember exactly when," Mahesh said slowly, coughing as he puffed at his *beedi* to light it, "but it's been some years. I remember it was the end of winter."

"Yes," said Jeevan, plucking a log from the pile of wood behind him and feeding it to the crackling flame that licked the sides of the saucepan. The distinctive smell of pine resin reached our nostrils, and people instinctively moved closer to the makeshift stove for warmth. "Back then the IVRI used to experiment on animals. Giving injections to cows and pigs, doing research on animal disease. Did you know?"

The question was directed at me and caught me by surprise. I was puzzled by the sudden mention of the IVRI, the Indian Veterinary Research Institute, established by the British in the nearby town of Mukteshwar in the early twentieth century. I wondered whether he had misunderstood my question about where the wild boar had come from. "I did know that," I said with some uncertainty, wondering how to turn the conversation back to boar.

"Yes, they only stopped when the government shifted the lab to Bareilly [a city in the lowlands of Uttar Pradesh]. But in those days they used to keep a lot of animals. It so happened that a pregnant sow ran away from the old cattle kraal because the caretaker had a habit of leaving the door open. I know because my father was a peon there then; they were all sent to look for her."

"Where did she run to?" I asked, intrigued by the direction the story was taking.

"The forest," said Mahesh, who had been nodding as Jeevan talked. His father had also worked at the IVRI. "The kraal was near the jungle, and she ran straight there. Maybe to have her children in an isolated place. They sent many men into the forest to look for her. Many villagers looked for her too. But she was never found."

"No, she never was," another farmer affirmed. "My father told us when he came home from work. He said that all the officials [bade sahib] couldn't find a pig between them. She taught them a good lesson." Laughter rang out at his mocking tone. "At the time, he said the leopard would finish the job (kaam tamam)."

My face must have betrayed my doubt because Jeevan took one look at it, and a defensive tone crept into his voice. "It's not a lie," he assured me.

> The pig ran into the jungle and was never found. A few years after that we began to see a lot of wild boar in the jungle. Nobody thought she would survive . . . it was a big jungle full of tigers and leopards. But she must have given birth to her children, and they gave birth to more children. Maybe some wild boar came from elsewhere and they had children together. Pigs have lots of children. . . . Now who knows what was in those injections they gave her. Maybe she had more children than normal. And I can tell you one thing, the pigs here have very large tusks. . . . All we know is that we never had such a problem with wild boar before that.

"Yes, the trouble started then . . . because that domestic pig went wild [soongar junglee ho gayee]," Mahesh concluded. "And now the state tells us that we cannot kill these pigs because they are wild animals. Now you tell us . . . are we mad or are they?"

*

The pig who went wild was a figure of legend in this region. I heard the story of her remarkable escape on several occasions in other villages that dotted the perimeter of the dense forest controlled by the IVRI. In some accounts, she did not escape to a life of freedom in the jungle but was forced by circumstances to live it. It was said that the person in charge of caring for the pigs, a Dalit man, would usually let them graze on their own by the edge of the forest. They would always return just in time for their evening rations. But one time, the pregnant pig did not come back with the rest. Perhaps she had given birth to piglets in the forest and couldn't leave them. But she was never found after that day. The entanglement of people's lives with pigs is thus viewed as an outcome of the distinctive history of an individual animal who linked people, animals, and institutions in unexpected ways.

The pig's story offers a glimpse into how "wildness" and the "wild"—as modes of perception, as states of being, and as forms of power and contestation—are produced in contingent and asymmetric circumstances. Protected by the power of the state and supposedly cyborged by science, the wilding of her body and those of her progeny illuminates the workings of power and hierarchy. Her story exposes the uneven and arbitrary nature of wildness, the colonial logics of which remain in force in postcolonial conservation laws in India. The Indian Wildlife (Protection) Act prohibits any hunting of wild animals except in special circumstances where an "animal has become dangerous to human life or to property (including standing crops on any land)." This, as I discuss later in the chapter, is in keeping with colonial exemptions to prohibitions on hunting in cases of damage to crops by wild animals. However, in practice, the postcolonial state, much like its colonial predecessor, rarely issues cultivators a license permitting them to kill wild animals who attack people or destroy crops, especially where agriculture is only for subsistence.

To add insult to injury, the wildness of the pigs was itself suspect. "And these bastards [wild boar] were domesticated [*paltu*] at some point. Ok, we can't kill leopards, that I understand. But pigs?" one man said incredulously. It was this disbelief at the protection of the wild boar that led Mahesh to declaim that if a state protected an animal simply on the grounds that it was "wild" when its wildness was so clearly historically contingent and unstable, then either the state had lost its mind or its citizens were losing theirs if they still believed in the possibility of seeking justice from the state. In this sense, the pig was a classic creature of empire, a postcolonial animal whose wilding was made possible by the biopolitical writ of the state that bestowed life on wild boar while dooming its human citizens to death.

Where does that leave "wildness"? Can it escape its "blighted colonial etymology" (Halberstam 2014, 140; cf. Nyong'o 2015)? Or is the runaway pig gone wild forever doomed to be relegated to the status of empire's foot soldier? Can her body be understood only as a site for colonial (re)production? In what follows, I argue that the pig's history, even as it bears indelible traces of racial meaning and the workings of sovereign colonial power, contains within it the potential for an *otherwild*, a messy wildness that reconfigures, unsettles, and exceeds the ways in which it is framed in projects of colonial and caste domination or in fantasies of human mastery of the nonhuman. Following Jack Halberstam, I imagine the otherwild as a space that is not entirely contained by the logics of rule and domination, whether by some humans over other humans or by some humans over nature. I do not mean to suggest that this otherwild is an unmediated and unregulated terrain of contestation or resistance. Instead, what I seek in the otherwildness of the runaway pig is the latent possibility of another world: a world of tentative and difficult fellowship, relatedness, and exchange; a world where animals are not always and already imbricated in human projects but come to interspecies relationship as beings whose histories, though linked to humans, are not exhaustively contained by them; a world where logics of domination and violence are remade in unfamiliar and potentially radical ways even as they are reinscribed.

Protecting Wilderness from Wildness: (Post)Colonial Racialization and Game Preservation

In 1904, the Government of India (GOI) solicited confidential comments from local governments on a draft it had prepared for a bill that would "preserve and protect" fish and game across the country.[1] G. Bower, the magistrate and collector of Saharanpur district in the United Provinces, dispatched a long response. He was, overall, in favor of increased protection for game, but he was disturbed by several clauses of the draft bill. In particular, he found the government's suggestion that the money obtained from the sale of hunting licenses to European and Indian sportsmen could be used to "provide guns gratis to those [cultivators] who hold licenses for the protection of crops or for the destruction of dangerous animals" galling.[2] In his dispatch, he bemoaned the difficulty colonial administrators already faced in establishing state control over cultivators who had ready access to cheap arms that were well-suited to the task of crop protection. The GOI's proposal to provide them with superior armaments, he warned, would create a situation in which men with "decent weapons" would roam the countryside unchecked, killing all the wild animals they came across in the guise of shooting dangerous animals. This, he noted wryly, would be a "very curious method of encouraging the preservation of game."[3]

His passionate response and those of several other officials from across the United Provinces and other provinces forced the GOI to change course and propose a simpler bill that did away with some of the offending clauses.[4] This series of exchanges reveal much about the complicated nature of colonial conservation. What stands out in the first draft of the bill are the conceptions of wildness and the wild that were in play at the time. Colonial officials believed that natives lacked self-control and could not be easily controlled by the colonial state either; their actions unfolded on an animal terrain that had to be mastered by colonial law.[5] Colonial accounts of native hunters (*shikaris*) abounded in stereotypes that cast them and natives more generally as wild, stealthy, and untamable by law and order, a construction that is echoed by a number of contemporary conservationists and forest officials.[6] In order to mark their racial distance from these men, upon whose skills they remained dependent, colonial officials and sportsmen laid emphasis not only on fair play but also on the professional nature of the *shikari's* work. White men, the narrative went, hunted for pleasure and to build character, but above all, they hunted to secure and bolster empire and establish control over life. The native, on the other hand, hunted for food and money. The need for subsistence shaped not just his hunting but also his character, making him deceitful and greedy and unwilling to adhere to the moral code of hunting articulated by the British. Through such discourses and practices, the native came to be positioned as the "anti-self" of the colonizer (Taussig 1987).[7]

What is also striking about this exchange is the GOI's attempt to balance its desire to protect game with its concern about the extensive damage caused by wild animals to cultivation. Game-protection laws were a constant source of disagreement between officials from the revenue and forest departments. Revenue officials were worried that these laws would interfere with cultivators' abilities to protect their crops, and they constantly reminded the government of the need to secure agrarian revenue alongside protecting game. There was thus a clear recognition that wildlife would have to be protected at the expense of cultivators, a calculus that remains important in contemporary India.

This tension between collecting revenue and protecting game only intensified over the next few decades. In the meantime, cultivators suffered significant losses from the depredation of wild animals, large animals whom they could not kill without a gun license. In 1925, during a legislative council session, the legislator G. B. Pant, a politician from Kumaon, noted that wild animals were causing significant damage to crops and cattle. "If statistics were collected," he said,

> the depredations of wild animals cost Kumaon about 36 lakhs in agriculture and livestock. This enormous wastage is largely due to the reservation of extensive tracks [as forest] and to the paucity of arms licenses. . . . The general policy of disarmament of the entire community followed side by side with reservation of extensive forests in the immediate vicinity of populated tracks is apt to raise a suspicion that the govt. cares more for the protection and preservation of wild animals than of their human neighbors.[8]

Pleas such as this did not always move colonial forest officials. While some were cognizant of the challenges of agrarian cultivation in these conditions and worried about the effect it would have on colonial revenues and rural unrest in general, a number of forest officials believed that the most important task that confronted them was the protection of wildlife from the wildness of the natives. As a number of *shikar* narratives and legal and public discussions around conservation reveal, it was around the same time that colonial discourses of conservation began to represent wild animals not as game to be preserved but as a scientific, cultural, and national resource in urgent need of protection. A number of colonial officials were quick to blame natives for the loss of wildlife. The only solution to the extermination of wildlife, they argued, was to entirely exclude natives (except the most elite, especially local princes and landlords) from hunting altogether. In the 1930s, conservation laws were deemed necessary in the face of the uncontrollable wildness of villagers, a wildness that made it impossible for them to understand the scientific, ecological, aesthetic, and cultural benefits of wilderness.

In many ways, these debates foreshadowed the tensions that continue to haunt conservation policy in contemporary India. As I mentioned in the previous chapter, farmers in Kumaon find it difficult to cultivate their fields in the face of daily onslaughts by wild animals who are now accustomed to living alongside humans, and many have chosen to abandon cultivation altogether.[9] Even though villagers can theoretically seek compensation from the state for damage to cultivation by wild animals, the process is difficult and expensive. To receive compensation, villagers have to file reports with both the forest and revenue departments, and these departments must conduct a joint inspection of the field before approving the claim. If such processes are slow around protected areas (Ogra 2009), they were positively glacial in the villages where I conducted fieldwork, which are not on the outskirts of officially designated sanctuaries or national parks. The offices of the Forest Department were also located at some distance from most hill villages. Most villagers could not af-

FIGURE 24. A field of potatoes dug up by wild boar.

ford to spare the time or the money for regular trips to these offices and would often have to plant a new crop in the fields that suffered damage before the inspection could take place. Either way, they preferred to just cut their losses and move on.

The forest department, though it was sympathetic to the plight of cultivators, continued to deny shooting licenses to villagers. In 2011, I spoke with a senior official from the forest department about the wild boar population explosion in the mountains. He agreed, in theory, that it was important to allow cultivators to defend their crops, especially from wild boar who were a "nuisance," but he pointed out that things were never that simple. "Let's say I issue a farmer a license to shoot wild boar," he explained to me. "Now what if he shoots some other animal, like deer or a leopard? I have no control over what he will shoot. But the conservationists will show up at my office and say that I didn't do enough to protect wild animals. Even if he is honest and shoots only wild boar, somebody will complain to the media that the forest department is allowing the slaughter of wild animals. Wild boar are protected under the Wildlife Act. So you have to be very careful." After some reflection, he added, "Often these villagers want to shoot boar because they like to eat them. The damage is in one place, but they will shoot in another. I can't sanction that."

I was struck by the subtle echoes of colonial racialization in his explanation for the forest department's denial of licenses to cultivators. While he recognized the scale of the damage caused by wild animals to cultivation and the difficulties of living alongside them, he was skeptical about issuing hunting licenses to villagers for reasons that were not very different from those of his predecessors—villagers were inclined to deceit and had little respect for the law. But his account of the difficulties involved in navigating this situation also raises another critical consideration: the power of wildlife conservationists and animal-rights activists who are dedicated to ensuring the conservation of wild animals at all costs. Even as the state and prominent conservationists grow increasingly warm to the inclusion of villagers in conservation initiatives, the latter in particular remain committed to the idea that wild animals are inviolate and must be protected at all costs, no matter the collateral damage.

Indeed, over the last decades, any attempts by the state to issue shooting licenses to cultivators besieged by wild animals have met with unfaltering and often angry opposition by animal-rights groups and some wildlife conservationists. This contemporary opposition also draws on older colonial narratives about the threat that native hunting poses for wild animals. Wildlife activists oppose farmers' requests for the right to kill predatory wild animals by claiming that the uncontrollable and wild nature of humans will only lead to the extinction of endangered wild animals. The natives are declared unworthy of trust; their wildness is capable of being controlled only through the imposition and strict enforcement of laws that enshrine the sanctity of wild animals who are represented as endangered or at risk of extinction. These modes of thought and practice are, to borrow Timothy Choy's (2011, 23) words, "environmental mobilizations of sentiment" around endangerment. Choy (2011, 26) notes that to speak of "endangerment"—which is at the heart of environmental politics worldwide—is to "speak of a form of life that threatens to become extinct in the near future; it is to raise the stakes in a controversy so that certain actions carry the consequences of destroying the possibilities of life's continued existence." In India, such affective discourses of endangerment are brought into play by a number of environmental activists to represent interventionist practices such as controlled culling as a violation of a pure and extraneous Nature.

For many villagers, what is frustrating about this fetishization of wilderness as autonomous and inviolate is the fact that it ignores how their lives have come to be related to those of wild boar not because of any conscious attempt on their part but as the outcome of a complicated colonial and postcolonial history. Having been figured for so long as a threat to the continued existence of wild animals, villagers find themselves unable to break with the violent forms of relatedness that have been thrust on them by this history. For better or worse, conservation laws have bound their futures to those of wild animals. Their situation bears remarkable similarity to that experienced by people in the Rajasthani village that the anthropologist Ann Gold worked in for many years. Gold (Gold and Gujar 1997, 78) eloquently describes how many villagers in Ghatiyali remember the "rapacious, insolent, unchecked power of" wild pigs in the past as being associated with the extractive and unjust nature of royal authority that shielded these boars and made their depredations possible. In Kumaon, too, the unbridled plunder committed by wild boar is linked to the whims of colonial and postcolonial states that value their animal subjects more than some of their human subjects.

In 2016, these debates about culling acquired new intensity when the Ministry of Environment, Forest and Climate Change announced that in view of the large-scale damage caused by wild boar in Uttarakhand, the animals would be classified as "vermin" in some districts across the state for a period of one year. As vermin, wild boar who wandered into agricultural areas could now be shot by villagers without prior permission from the forest department. Villagers I spoke with in the wake of the order were, however, skeptical. They pointed out that guns and ammunition were not easy to acquire given the district administration's general reluctance to issue firearm licenses. Putting out poisoned food, farmers said, ran the risk of affecting animals other than wild boar who were still protected under the Wildlife Act. All told, they were dubious that the order would lead to any change in their relationship with wild boar. In any case, this speculation was rendered moot by people's realization that the hill blocks in the district that I worked in were not subject to the new regulation, which applied mostly in the plains districts of Uttarakhand.

At the national and regional level, however, fears about marauding villagers in pursuit of wild animals were rampant. Members of the National Board for Wildlife expressed dismay at the "retrograde" decision and worried about the absence of checks on "who [would] kill."[10] Some forest officers in Uttarakhand privately shared this concern and suggested that villagers would kill indiscriminately for meat. Meanwhile, Maneka Gandhi accused her colleague, the environment minister, of displaying what she called "a lust for killing animals." What the environment minister described as "scientific management" of wildlife populations, she termed a state-sanctioned "massacre" of animals by villagers. Gauri Maulekhi declared that the state was using the "Wildlife Protection Act, formed to protect animals, to hunt them." "If human-animal conflict was a concern," she commented, "why has the chief wildlife warden taken no steps to remove encroachment from inside and on the border of the forest and preserve the wildlife habitat?"[11]

Through such colonial narratives, wild animals were fixed within the realm of Nature, and came to be defined as those who live an "authentically natural life, untainted by human contact" (Ingold 2000). The boundary between "wild" and "domestic" was fixed too, and its displacement was understood by people such as Maulekhi in terms of an "encroachment" by humans who are "out-of-place" (Philo and Wilbert 2000). There was little room in this understanding of wilderness for precarious histories of wilding such as that experienced by the runaway sow. But the story of the runaway pig reminds us that there is no pure nature that is not always and already mediated by history and politics. The particular trajectory of her history demands an exploration of the errant and unpredictable nature(s) of animals that cannot be contained and comprehended within the ambit of human intentionality even as their trajectories are entwined with humans. In the next section I follow the lives of pigs as they go wild (and domestic), tracing the ecology of practices that undergird this wilding and produce an entanglement of human and nonhuman fates.

Paltu-junglis: Understanding Weedy Wildness

Wild boar (*Sus scrofa cristatus*) look like they are built for battle. They are squat and muscular and have thin, patchy fur except for the luxuriant upright mane of bristles on their neck and back that stiffen and quiver when they are angry. When they feel threatened, they charge the threat and slash it with their tusks, protruding teeth that are bigger on males than females. This is the *tearing* of the body that villagers describe, an especially evocative word that foregrounds the fragility of human bodies and the ease with which adult boar can take them apart. Adult sows have sharp canines and are not afraid to use them, especially when they have litters of young with them. Wild boar, like domestic pigs, are prolific breeders, and can have up to ten piglets at a time if the conditions are right.

People in the mountains often told me that they were more afraid of wild boar than leopards. "If you come across a leopard, chances are that it will run away. Unless it's a man-eater, of course," an old man once told me at the start of my fieldwork when cautioning me to watch out for certain wild animals. "But if you have the misfortune to run into wild boar, then you should immediately climb a tree. Or they will definitely tear you open." When I finally encountered a group of wild boar a few weeks later, I was on my own, standing in a terraced field farther up the mountain from them. What first got my attention was the sound of grunts, which sounded louder than they were in the midday silence. When I looked around for the source of the sounds, I spotted several dark-gray bodies standing in the middle of a small brown patch that stood out in the otherwise shockingly green landscape. They had rooted up and consumed the crops on the narrow field on which they were standing with only a small patch of pea plants remaining for them to eat their way through. When they looked up a few minutes later and saw me standing on the horizon, one of the boar stopped eating the delicate white pea flowers that she had been savoring until then. She grunted and held her body taut, the muscles in her back stiff with anticipation. The other boar looked at her every now and then but continued eating. After a few moments, when I shifted my weight, the boar grunted again and charged a few steps forward, butting the wall of the terrace. Unsure of my ability to outrun her in case she did charge me in earnest, I turned tail and ran toward the safety of the nearest house.

"The problem," Leela, a sympathetic neighbor said as I recounted the story of my escape over tea in her kitchen, "is that these pigs are just like domestic pigs. They keep hanging around here." Tendrils of smoke curled around her face as she bent to blow vigorously on the fire smoldering reluctantly in the mud stove, the logs of oak and pine having been drenched by the day's rain. "If they're wild then they should live in the forest, eat wild plants. But they want the best food—wheat, potatoes, peas. If an outsider saw this, he would say that we are raising them."

Her husband, Mohan, who was sitting in the front room of the house and watching television while listening distractedly to our conversation, perked up at this. "Yes, that's absolutely right," he said. The blue walls behind him were radiant with the reflection of the television, moving shadows mirroring each flicker on the screen. "The forest department has dumped the responsibility for these pigs on us . . . Now what should we call them? Domestic pigs, wild pigs, who knows," he spat out sarcastically before returning his attention to the devotional program he was watching. After a moment, however, he had an epiphany and walked into the kitchen. "Think of them," he said, adjusting his weight on a small *chowki* (wooden stool), "as *paltu-jungli* (domestic-wild). There's something wild in them, that much we admit. But all pigs have that compulsion in them, that much I can tell you. Have you ever heard of the sow who ran away from IVRI?" When I said I had, he smiled.

"*Yeh,*" he said, bringing his hand down in the air delightedly, knowing that I had understood his point without the need for more explanation. "Look at that. She wanted wildness [*junglipana*], and she ran away. All of them have that compulsion. But even if they are wild, they can't stay without us. They must have a relationship with us. Look at those wild pigs outside. That's why I'm saying, they are *paltu jungli.*"

Mohan's wonderful neologism, *paltu-jungli*, or the domestic wild, captures the fluid and contingent relationship between the wild and the domestic perfectly. In effect, he was arguing that *all* pigs are always and already subject to the affective tension between the wild and the domestic. Pigs who are wild consciously seek out humans. But even pigs who are in domestic relationships with humans long to escape the confines of that relationship and go wild. The story of the runaway sow exemplified this compulsive yearning that pigs feel for the wild. The sow kept by the IVRI could no longer ignore the siren song of wildness; she *wanted* wildness and she ran away in search of it. For Mohan, wildness was best understood not as a stable state of being but as a tendency, a *desire* that is latent in all (porcine) beings. Wildness thus takes shape as an excess of desire that, despite the best efforts of humans to corral it, can sometimes spill over in ways that they do not anticipate or understand and cannot control.

I was reminded of the difficulty of establishing human control over porcine inclinations for wildness in 2011 during a visit to the town of Almora, when I met Prema, a Balmiki woman who kept several pigs. Pigs, she said, were easier to care for than cows and goats. She would let them out of their pen, a rickety structure on the road right under her own home, in the morning. They spent most of the day eating a mixture of grass, trash, and small rodents. In the heat of the day, they would lie on their sides in the cool ditch by the side of the road. When the sun finally lost some of its heat, they would come home, coated in mud, and lie outside their pen. But one pig, she complained, had been giving her a lot of trouble of late. She pointed the pig out to me from where we were sitting in her courtyard. He was lying on his side, and I could see that his bristles were coated in a dark-gray sludge; underneath the sludge, I glimpsed faint patches of pink. He was an impressively large creature. The several piglets who were squealing and grunting their excitement as they chased each other in the waning light of the evening gave the big boar a wide berth. If they got too close, he would snort a loud warning without moving his body, causing the piglets to retreat and seek protection with their mothers.

"He's developed a terrible habit," Prema grumbled as she picked small stones out of the grains of uncooked rice she had in front of her. "He disappears every couple of days . . . doesn't come back at night. I worry that a leopard will kill him, although he is so big that a leopard would be mad to try and attack him." I was intrigued by this pig's regular disappearances, and asked where he would go. "Someone told me that he has an *adda* (a lair) in the forest," Prema told me. "I don't know what he does there. He comes back after three or four days, but I'm afraid that one day he just won't come back."

Prema recognized that her pig *chose* to exist in a somewhat fluid state of domestication. She was unsure how long he would be content with only two or three days in the forest. She told me that she had beaten him with a stick when he came home after his first few disappearances. She had even tried locking him in the pen for a few days. But he was incorrigible, and she finally gave up. She could not change his ways, she told me, but attempted to establish a bond between them through other means. She offered him potatoes, one of his favorite foods, every time he returned home from the forest. In fact, that evening she had already kept aside a couple of boiled potatoes to offer him as a reward for his homecoming. "Cows and goats are not like this," she observed; "whatever happens, they come home at night. But pigs are half-wild . . . All I can do is give him the potatoes. This way, at least a relationship stays in place." Prema's boar shared a relationship with her that was, as I said in the introduction, domestic but not domesticated. He was half-wild, she had said, and disinclined to give in to her efforts to discipline him. As I watched Prema make her offering to the boar before prodding him to enter the pen with the other pigs, a request that he granted with the greatest of reluctance, I thought to myself that potatoes were a rather tenuous substance to base a domestic partnership on. Yet on further reflection, perhaps they were as good a mediator as any.

In noting the inexorable tendency toward wildness in pigs, both Prema and Mohan were attesting to the common wisdom—not to mention the observations of biologists, historians, and ethologists—that pigs are inclined to go feral. In the United States of America, for instance, the population of feral hogs has exploded in the past few decades and with it, state costs of surveillance, disease monitoring, control, and culling. The damage they do to crops runs into billions of dollars. Much of this population originated as domestic swine brought over by settler colonists. In *Creatures of Empire*, Virginia DeJohn Anderson (2004) offers a glimpse into the history of their wilding. She describes how English colonists in America initially sought to recreate patterns of English civility through their relationships with their livestock animals. However, they soon diverted most of their labor to the precious tobacco crop, neglecting their animals and letting them roam free. "Freed from many of the restraints of English style husbandry, livestock changed too. . . . Unruly swine grew fiercer and roamed the woods in formidable 'wild gangs'" (Anderson 2004, 108). These early gangs of escapees were later joined by European wild boar (*Sus scrofa*) that were initially brought over by colonists in large numbers in the eighteenth and nineteenth centuries to provide sport for hunters and that subsequently fled the confines of game reserves. These two populations soon interbred to produce the feral hog that roams across most states in North America today.

It is not difficult to imagine a similar history for the runaway sow. Perhaps she gave birth to her piglets in the forest and couldn't leave them to go back to the kraal. Or maybe she found life in the forest preferable to being penned up in the evenings and decided that she didn't want to return to her captors. The guarantee of food and shelter may not have been as tantalizing as the pleasure of freedom. It is possible that she and her piglets bred with wild boar that were already living in these forests or had moved into them from adjoining stretches of forest. The mixed oak forests in this region are rich with resources that could have easily sustained a domestic pig and her progeny. She might have munched on the acorns and other forest floor products that abound or on the luxuriant fronds of grass that cover the slopes in the monsoon. She could have had her pick of insects and reptiles. It would not have taken long for her and her descendants to transform morphologically over the course of this process, especially if they were breeding with wild boar. Over time, as they adapted to new environmental and genetic conditions, their heads would have grown larger and they would have started to develop a dorsal mane.

While this history may be difficult to confirm without genetic tests of the wild boar population in this region, the story of the runaway sow resonates with what we know about the tendencies of domestic pigs to go feral and bears strong resemblance to accounts of the wilding of domestic animals from other spatial and temporal contexts. Like the stories about outsider monkeys, the tale of the pig gone wild certainly could be described as conspiracy theory inasmuch as it is a narrative about animals that is "constitutive of critical social commentary" (Mathur 2015), but it acquired credence and force from what people know about the inclinations and dispositions of the animals in question.

For villagers, there was little doubt that some of the wild boar who shared the landscape with them were descended from the runaway sow. The boar, they said, have a fondness for lingering in the village that can only be explained by their domestic origins. Several farmers complained that wild boar were no longer content with just raiding crops and returning to the forest but spent all day lounging in fields! The adult pigs slept for hours in large troughs they dug in the soil. Often they had piglets with them, which made approaching the area a dangerous proposition.

The humans who had to learn how to share space with these pigs devised ingenious strategies to navigate this difficult copresence. Dinesh, an eighteen-year-old boy who had a reputation for being too fond of the girls, once told me and a group of others that he had a foolproof plan for making it past a group of wild boar without injury. "Have you heard 'Main Tera Boyfriend, Tu Meri Girlfriend'?" he asked us. Everyone nodded. The song was all the rage among village youths in the summer of 2015. Strains of the catchy refrain *na, na, na, na* would waft over on the breeze as people walked by listening to it on their cell phones. "I was walking up to the main road one time and listening to this song. Suddenly I heard these noises." He started to grunt vigorously.

"I thought one of my friends had played a prank on me by recording some new joke version of the song on my phone. But then I saw some wild boar running away. Maybe they don't like the song. Like my *chacha* (uncle)." He gave me a look full of meaning. His uncle had smacked him a week ago in my presence for playing "vulgar" songs. "Now I don't walk anywhere without playing the song. Let's just hope the pigs don't grow to like it. Otherwise they will also show up to dance. And I want a prettier girlfriend," he said with a straight face as everyone else collapsed in helpless laughter.[12]

Others, understandably, did not face the prospect of this copresence with such equanimity and good humor. People were still surprised by how accustomed the wild boar were growing to living alongside humans and unsure of how to make sense of the resultant interspecies intimacy. "They look like the domestic pigs you see sleeping in the ditches near Almora . . . look how they just lie here without any fear" Rachna *mami* in Loshi told me. "Some people from Delhi were walking around here the other day. The lady asked us if we kept pigs because she had seen some in the fields below the road. I laughed and told them that these were dangerous wild animals. You should have seen their face." I could not help but chuckle at the image. "But I also feel scared now," she continued. "What if the pigs had attacked these tourists? They didn't even know the animals were wild. We will have to put up signs now," she said, only half-jokingly. "These pigs are not domestic," she intoned, sketching a signboard in the air with her fingers.

In noting that it would now be necessary to declare that wild boar were not domestic, Rachna *mami* was, like Mohan and others, pointing to the ways in which these animals constantly subverted the boundary between wild and domestic through their everyday actions. In many parts of the world, wild boar act as a "weed species" (Richard et al. 1989) because, like weed plants, they seek out and thrive in contexts that are often characterized as "human dominated." However, the term *human dominated* is a misnomer because it conceals how the landscapes categorized as such are actually created through the labor of shifting and contingent assemblages of interdependent humans and nonhumans. Perhaps it would be better to say, then, that wild boar world themselves into ongoing multispecies worlds. In Kumaon, wild boar prosper as they consume the fruits of the labor performed by multispecies assemblages of which humans are one, important, part. In general, wild boar are omnivorous and eat grains, vegetables, grasses, fruits, tree bark, carrion, reptiles, insects, and even mud. The animals I encountered in Kumaon, however, displayed a particular fondness for agrarian produce, squealing and grunting with delight each time they unearthed a potato, peanuts, green shoots of rice, or a cluster of kidney beans.

FIGURE 25. Maize damaged by wild boar. Such damage is so routine that villagers do not bother filing for compensation.

I witnessed the discerning palate of wild boar on several occasions, but one stands out in my memory. As dusk fell over the valley one evening, I was washing my dinner plate when I spotted movement out of the corner of my eye. I turned my head and saw a male boar with large tusks wander into a nearby field. He began to dig frantically in the soil until he found the new potatoes, whose smell had clearly drawn him to that field over the others planted with peas that he had walked through without a second glance. Where he had been frenzied before, he was now unperturbed, savoring each potato he excavated with visible enjoyment. This was food to be enjoyed, not just consumed for the sake of eating. Every now and then he would roll around in the mud before getting up and starting to dig again. He stayed in the field for an hour, grunting softly each time he unearthed another potato. As I watched him, I remembered that numerous people had told me that wild boar are gourmets, always going for the *mandua* (finger millet) before the wheat, the potatoes before the onions, the peas over the chives. I now understood what they meant.

Their clear preference for some crops over others, however, did not get in the way of the rapidity and thoroughness with which wild boar ate their way through agrarian produce. Seeds sown on one evening were often gone the next morning, leaving people angry and heartbroken over their wasted labor. What the boar didn't eat, they uprooted in search of *kurmula*, a white grub that they relished. When they left after having eaten their fill, the only traces of their presence were the telltale untidy furrows in the soil. Wild boar dependence on agricultural crops, especially in areas at the interface of fields and forests, is well documented not only in India but in many other contexts as well.[13] A study of damage to agriculture by wild pigs across eleven states in India found that the "damage to agricultural crops was enormous and wide-spread. [Wild boar] fed on all phenological stages, but tender stages and matured crops were highly susceptible to damage" (Chauhan 2011).

The weedy nature of multispecies worlding was semantically evoked in the local term used to refer to wild boar, which was *soongar* (pig). People would sometimes employ the adjective *jungli* (wild) to indicate that they were talking about wild boar, but more often the term they used was just *soongar*. The unqualified use of the noun, I believe, signals something important—the instability of wildness and domesticity as categories and states of being, an instability that emerged out of the tight interconnections that produce multispecies worlds. Pigs, by their very nature, lead lives that are simultaneously entangled with and yet unrestrained by humans. Rather than making them taboo, as Mary Douglas has argued, the liminal position of pigs in this case frees them from the confines of a colonial wildness, allowing them to move deftly in and out of multiple, overlapping worlds. This liminal status is perfectly summed up by Mohan's term *paltu-jungli*. They seek immersion in human spaces but also claim freedom from human expectation. Even though domestication relies on an interimplication of human, animal, and plant lives, it is a fragile and constantly shifting relationship that is characterized more by hope, faith, and daily effort than by any certainty of power.[14] The irrepressible tendency of pigs toward wildness should not be read as a failure of human control, then, but as a natural outcome of the vibrancy and liveliness of pig mind and flesh. Prema's lament that she did not *know* what her pig did in the jungle is a powerful reminder of the persistence and importance of difference even in the most intimate interspecies relationships. These interactions are lessons in how to live with difference. The *otherwild*, then, is perhaps best sought in this space of unmasterable difference, in the humbling recognition that animal lives, even as they are coconstituted alongside human lives, exceed their imbrication in the latter.

Jungli soongar, jungli log: Wildness across Species

Pigs are semiotically charged creatures whose bodies accrete layers of sedimented meaning. Across much of India, pig rearing and the consumption of pork is associated with Dalit and tribal groups. In Uttarakhand, upper-caste Hindus, but also a number of *thul* (literally big, meaning higher status) Dalit *jatis*, despise pigs and describe them as unclean animals who enjoy eating excrement. Pig rearing is shunned by upper castes, and the very sight of domestic pigs immediately signals the presence of lower castes to them. Those who eat them or are rumored to do so, mostly people from the lowest *jatis*, are singled out for caste humiliation and are described by upper castes as ritually impure. In December 2012, I witnessed Manish, a young Pundit man, spit repeatedly out of the window of the shared taxi we were traveling in when we passed a group of pigs wallowing in a ditch right on the outskirts of Almora. Pigs were an unusual sight in villages; indeed, hardly anyone in rural areas reared them. Manish's uncontrollable spitting signaled his visceral disgust at the sudden sight of these pigs, a feeling of physical recoil so strong that he asked the car to stop before retching violently at the side of the road. When I asked him what was wrong, he said, "yeh bhangi ilaaka hai" (this is a *bhangi* neighborhood)." The term *"bhangi"* is a pejorative term of abuse for Balmikis, a formerly untouchable group at the bottom of the caste hierarchy because of their traditional association with the cleaning of human excreta. When I protested his use of the term, he was firm. "You have no idea," he said to me, "you're from the plains. It's different there." I was about to point out that caste discrimination was as entrenched in the lowlands as it was in the mountains, but he kept going before I could say anything.

"These people keep pigs. *Chhee*, what dirty animals. Only dirty people could keep them. Do you know how they kill them?" His description of how "these people" would supposedly lock pigs in a room with holes in the walls and then repeatedly insert iron rods through these holes until the pigs were bleeding and could be easily slaughtered was so graphic that another woman in the car begged him to stop. Throughout this narrative, I kept glancing at Chandan Arya, my Shilpakar neighbor, whom I was accompanying on a visit to Almora to see his granddaughter. Chandan *da* and his extended family had become close friends, and I knew that he resented the subtle ways in which caste power was exercised by upper-caste villagers. I wondered how he would react to Manish's assertion of caste privilege through this open expression of *ghrina* (visceral disgust).[15] When I tried to counter Manish's claims a second time, he chuckled at what he read as my naïveté. "Even *Harijans* agree that these people are dirty," he said. "Isn't that so, Chandan *da*?" Chandan da nodded noncommittally.[16]

"See," said Manish, turning to me. "Those people are *jungli* [wild]. When pigs die, human shit comes out of their stomach." The Pundit woman who had begged Manish to stop describing the killing of pigs closed her eyes and spat a thick gob of saliva out of the window. "And they eat that shit. Only wild people would do that. I wouldn't touch someone like that, doesn't matter what their caste is." Later, when our shared taxi finally arrived in Almora, he took me aside to warn me against eating anything in Chandan *da*'s sister's house. "They're not *bhangis*," he said, "but all Harijans like to eat pig meat. And you will be sick if you even smell it. They're dirty people. Nothing will happen to them." When I stiffened with visible anger and told him that I would eat whatever I wanted and wherever I wanted, he shrugged his shoulders and left for the bazaar.

Even though Manish claimed that his disgust had nothing to do with caste identity and was directed only at dietary practices, food and caste in South Asia are inextricably connected. Upper-caste hierarchies of ritual purity and pollution are enforced through taboos on the consumption of food not only in lower-caste households but even in the presence of lower-caste people. Food taboos, on the face of it, have relaxed in recent times. It is not uncommon to see upper- and lower-caste villagers, usually men, eating together at restaurants. Some lower-caste villagers pointed to the fact that they were now served tea in upper-caste homes as a sign of how caste discrimination was less rigid than before. However, caste distinction and hierarchy, though less explicit, are still enforced through outwardly "neutral" forms of speech and action (Pinto 2006). If a lower-caste villager offered an upper-caste person a glass of water, the latter would politely decline by saying that they had just drunk some even if they were parched with thirst. Both knew that the reply, even though it was phrased innocently, was loaded with caste privilege and disgust. Food, too, was similarly declined.

The idea that lower castes such as the Balmikis, who, in Uttarakhand, lived primarily in small mountain towns, would eat *anything*, including human shit found in the stomach of dead pigs, was central to upper-caste representations of them as beyond the pale of caste society. Manish called them *junglee*, a term that other upper castes also used to designate lower-caste people as irredeemable others of civilization and progress. They were, in his view, *neech log* (low people) prone to a certain kind of wildness. This wildness, he would argue, was manifest in their inability to curb their base instincts, such as the desire for pig flesh, despite the fact that these instincts violated the basic demands of health, hygiene, and civility. However, his ostensible concerns about hygiene and health were, at their core, articulations of caste disgust. Those less explicit than Mahesh would say that they had nothing against lower castes but could not understand why they would insist on consuming the filthiest animals. "Yeh jaat ki baat nahi hai, yeh safai ka mamla hai" (this is not a matter of caste, but a concern for cleanliness), one upper-caste man told me when trying to explain why he thought eating pork was so disgusting. However, as scholars of South Asia note, seemingly value-neutral talk about dirtiness and cleanliness is actually a coded way to reference caste and to "whis *achut* [untouchable]" (Pinto 2006). Ideas about dirt and cleanliness, as Mary Douglas (1966) so powerfully pointed out, symbolize deeper concerns about ritual order and disorder and allow upper castes to talk about caste without really talking about it. It is in this context that pigs become such an important symbol of caste difference.

How is the runaway sow positioned relative to these languages of hygiene, health, and modes of consumption that reproduce caste difference and discrimination in coded form? The story of her wilding, I found, opened up a space for lower-caste villagers to articulate subtle critiques of upper-caste othering of them even as it reinscribed caste(ist) notions of the transactional nature of personhood. I was, initially, surprised to learn that the same upper-caste people who expressed such visceral disgust at the thought of eating pork considered the meat of wild boar a delicacy. Even Pundit women who proudly proclaimed their vegetarianism would make an exception for the rare feast that followed the "poaching" of a wild boar. However, they did not view this as a contradiction. When I asked one woman what distinguished the meat of wild boar from that of domestic pigs, she claimed that all wild animals ate nutritious food in the forest, not shit and garbage like domestic pigs. Someone else joked that he was confident that any wild boar shot in his fields would be "tasty" because of their diet of corn, potatoes, peas, and maize.

On one evening in 2011, I was sitting in the stone courtyard of the house of a Thakur man named Kishan Bisht. Earlier in the day, his wife had invited me to their home to see their cat's newborn kittens. I sat on a striped cotton rug, shifting constantly as the heat of the sun-warmed stones burned through the mat and my clothes. The kitten asleep in my lap mewled in protest at my movement. I listened to the conversation that was taking place next to me. Kishan was asking Dhani Ram, a Dalit man who lived in another part of the village, whether he could borrow his gun to hunt wild boar. "I really want to hunt a *soongar*," he confessed. "It's been too long since I had their meat. I'll settle things with the local forest guard. He owes me a favor. Don't worry. Nobody will find out." Dhani was clearly reluctant to do what Kishan was asking. If things went wrong, he would be in serious trouble and could even be sent to prison. In the end, he told Kishan that his gun was not functioning properly. It was an old piece, belonging to Dhani's grandfather, and could easily misfire. Kishan was swayed by the wisdom of this and decided to put his hunt off until he could find a firearm.

As we walked back together in the direction of our respective homes, I said to Dhani Ram that I thought he had been put in a very difficult situation. He agreed. He was upset that Kishan had asked him for the loan of his gun. "These upper-castes are very two faced," he said, his anger evident in his words. "You have been here for some time now. You tell me. They say that people who eat pigs are *neech*. They spit on them. But they themselves lust after these *jungli soongar*. Now who knows whether these *soongar* are really wild or just domestic pigs who have become wild. How do they consider themselves higher [in the caste hierarchy] than us?"

On another occasion, in 2014, two years after Manish had reacted with such visceral repugnance to the sight of pigs in Almora, Chandan *da* and I recalled the events of that day over a cup of tea. "The most amazing thing," Chandan *da* said to me after recalling the numerous occasions that Manish had expressed his caste arrogance (*ghamand*) in the village, "is that you offer that man [Manish] any meat saying that it is wild boar, and he will eat it. It could be the flesh of a domestic pig and he will praise it." When I asked him if he ate wild boar, he firmly said "no." "Look, you know the reason we have so many wild boar in these forests is because one pregnant sow ran into the jungles of the IVRI and had hundreds of babies. Now to eat the children of a domestic pig and say that it's all right because they are wild . . . I don't think it's right."

Chandan da's comments, which I sometimes heard echoed by other *thul jati* Shilpakars in the village, reinscribed Brahminical notions of cleanliness and filth that undergirded caste hierarchies. Neither Chandan Arya nor Dhani Ram were willing to question the idea that the consumption of pork was ritually polluting. In fact, Chandan *da* was quick to note that he did not eat pork and would not consume the meat of a wild boar either. Those who did eat pork, he said, were *neech*, a statement that was clearly shaped by upper-caste ideas about caste hierarchy and food taboos. However, there was also, in these declarations, a critique of upper-caste power and its hypocrisy. Talk of the unstable and contingent wildness of the descendants of the runaway sow provided a counterweight, though subtle and not immune to its own forms of disempowerment, to caste subordination. Chandan *da*'s wicked joy at the thought that Manish was separated from the Balmikis, who so disgusted him, only by the thin and tenuous thread of wildness embodied by the pig who went wild speaks to how the fluidity of wildness, as idea and experience, can open up spaces of critical potential even as it consolidates and reimagines relations of domination. The momentary slippage of domination as wildness follows its messy and unstable course produces the otherwild as a fragile site of maneuver and negotiation where the carefully constructed structure of caste alterity and domination is challenged by the antiself of the upper castes. It is through inhabiting this otherwild that lower castes are able to expose the

FIGURE 26. The Indian Veterinary Research Institute campus is dotted with colonial-era ruins. This is one of their animal shelters.

brittleness of the boundary that separates their wildness from upper-caste civility.

Debates over what wildness means in light of the history of the escapee sow create otherwild spaces not just for questioning caste domination and subordination but also for critiquing the state for its neglect of *paharis*. The runaway sow shares a complicated relationship with the state. On the one hand, the narrative of her escape constitutes a blow to the power and control of the state and its institutions, particularly the Indian Veterinary Research Institute. Villagers have mixed opinions about the IVRI. Despite its downsizing in recent years, it remains one of the largest employers for about a hundred miles in each direction, maintaining a large staff of workers to care for animals and perform other tasks associated with the day-to-day running of the institute. Most families in these villages have at least one member who works at the institute or has worked there in the past. At the same time, villagers resent its control over a large section of forest land, a legacy of the colonial period when it was given charge of this three-thousand-acre forest to ensure its fodder needs. As is the case in other reserved forests, hunting is strictly prohibited as is the cutting and collection of firewood. Village women, in particular, resented this injunction and often sneaked into these forests to cut and collect wood when they knew the forest guard was elsewhere.

The pig's escape from this imperial institution was a source of great mirth to villagers. On multiple occasions I heard how the officials at the institute had been unable to locate the pig despite their resources. The fact that she had made her escape from the old cattle kraal was, in itself, of symbolic significance. Today, the kraal is what Ann Stoler might call a "ruin of empire," an abandoned building bearing a sign that it was built in 1903. Even though it is no longer in use, it nonetheless visibly and viscerally "reactivates" the "effects of empire" (Stoler 2013). Back in its day, however, it used to house the animals of the IVRI, its position atop a hill making it visible to villagers for miles. It was from this citadel of empire that the runaway sow took flight. Given that colonial science often reinscribed social and racial hierarchies, it is no surprise that villagers take such delight in the fact that the wilding of the sow took place against the will of a powerful institution of the state,

However, the sow and her descendants cannot simply be celebrated as rebels against the state. Villagers argue that the tenuous wildness of these boar is protected by a state that cares more for animals than it does people. This wildness is fostered not just through the strict wildlife conservation laws discussed earlier in the chapter but also through the state's neglect of agrarian and rural development. What allows pigs, like the outsiders discussed in the previous chapter, to march through this landscape and occupy it, villagers say, is the emptying of this landscape by young men and women who are no longer committed to a life in agriculture (Govindrajan 2015c). Many villagers lay the blame squarely at the doorstep of the state. One young man recounted the experience of trying to get compensation for the damage done to his crop by wild boar. "On my fifth visit to try and convince them to inspect my field," he said, "I gave up. Who has that much time and money. I'm certainly not getting any money from cultivation." He was bleak about the prospect of staying in agriculture. "What kind of life would it be if I grew maize and potatoes only so that wild boar could eat them? The government doesn't compensate our losses. They don't let us shoot these animals. And there are no new schemes for agriculture. So why bother spending every night sitting in my fields with crackers to scare away wild boar?" His feelings were echoed by many others, signaling a widespread disenchantment with agriculture and the state.

Some people even suspected that just as the land mafia had dumped monkeys on villages, the state, too, was deliberately unleashing wild boar on villagers to force them to abandon agriculture and sell their land to outsiders. It was rumored that local officials were keen to encourage the sale of land so that they could profit from the cuts that buyers would give them. This mistrust, born from a sense of desperation, was most poignantly reflected in a story that many villagers told me over the course of my fieldwork. Sometime in the first decade of the twenty-first century, a *pukka* road had been constructed linking a set of villages that ringed the forests controlled by the IVRI. This road had broken down in the very first monsoon that hit the region after its construction. This was not a surprise to people who were accustomed to the lightning-quick degradation of roads during the monsoon. After all, stories of contractors who became fabulously wealthy by skimming off the surface of road-building contracts were legend. Unlike other stretches of road, however, this road was never repaired. "You see, the road was only of use to us poor villagers," one man said bitterly. "No tourists drove on it, no trucks full of stone for building projects. So why would the state bother to repair it?" I was told by numerous people that the road was now the domain of wild boar who used it to travel between villages and the forest. "You can't walk on the road alone, even in the day," one woman warned me. "There are always wild boar to be found on it. We go in big groups whenever we walk on the road."

The story of the pig who went wild provided an opportunity for people to mount a critique of a state whose beneficial effects they found largely absent in their lives. Some people marveled at the fact that the state was so in thrall to these animals that it was building roads for them. One man wondered whether houses would follow. After all, it was too much to expect these animals to sleep out in the open. Perhaps people would be asked to trade places with them. The state was mocked savagely in these narratives. The joke, some people said, was on the state for placing its resources in the service of descendants of a domestic pig. Or maybe, said others sadly, they were themselves the butt of the joke, poor people who were less valuable to the state than pigs who now traveled on roads.

Otherwild: The Semiotic-Material Mess of Wildness

Let me return, then, to what the story of the runaway sow illuminates about the nature of wildness. The notion of "wildness," as a number of scholars have pointed out, is deeply entwined with colonial formations of race, sexuality, and gender. Colonial conceptions of wildness and the wild were employed with a view to deepening power and domination. The native was a figure who embodied wildness, fearful and seductive. Simultaneous to these other constructions, wilderness was also imagined as a pristine, empty space populated only by wild plants and animals untainted by human influence, a romantic fiction made real by the marginalization and expropriation of native people.[17]

In contemporary Uttarakhand, as across India and other parts of the world, the postcolonial state's fixing of certain spaces and animals as wild and their subsequent "protection" from unruly citizens is a process that is driven in complex ways by these colonial imaginaries. The belief that villagers cannot be trusted to defend themselves against predatory wild boar and that their innate wildness compels them to destroy precious wilderness is a legacy of colonial racialization. Conceptions of wildness, as we have seen, are also at the heart of relations of caste domination and subordination. Upper-caste villagers' attempts to establish distance from people of "low" *jatis* through a politics of disgust relies on powerful constructions of the latter as "wild" and beyond the pale of civilized society.

The question that remains is whether wildness is entirely circumscribed within these violent histories of differentiation and exclusion or whether there exists the possibility of an otherwild, a space and experience of wildness that remains infused with power even as it exceeds the ways in which it is imagined and deployed in colonial projects of rule and in imaginaries of human mastery of the nonhuman; a wildness that is *not* "incessantly recruited by the needs of order . . . so that it can serve order as a counterimage" but that contains within it the possibility of questioning established structures of domination (Taussig 1987, 220). This is a wildness whose force, autonomy, and effects are an outcome of its own vital potentiality and not just human representation.

The wildness of the runaway sow, I have argued, embodies this *otherwild* in that it is not easily contained by logics of racialization and rule. Her wildness was semiotically and materially messy, shaped as much by the experience of disorder and unruliness as it was by attempts to impose order. Her body and the bodies of her progeny, protected as they were by law, encoded the rule of the state. However, her escape from captivity by an institution of the very same state exposed the limits of its control. This was wildness run awry, and it was a powerful reminder of the effective materiality of nonhumans and the impossibility of total human mastery over them.

The messy trajectory of the sow's wilding also destabilized forms of differentiation that were underwritten by conceptions of wildness that replicated colonial racial constructions. The obvious contingency of her wildness provoked reflection on the instability of categorizations of certain groups of people as wild. People designated as belonging to a *neech jati* (low caste) would, on rare but important occasions, invoke the story of the pig who went wild and the lesson it provided on the fluid relationship between categories such as wild and domestic in order to challenge the fragile boundary between their wildness and the supposed civility of upper-caste villagers. In another realm, the protection and preservation of this unstable wildness in conservation law conjured mockery of and anger at the same state's bestowal of such benefits on pigs and neglect of its people.

Finally, the radical potential of the otherwild lies in its promise of a world in which interspecies relationality, entanglement, and involvement do not erode difference but emerge from a recognition of it. When recounting the story of the wilding of the pregnant sow, villagers would note that it was impossible to curtail the impulse to wildness that lurked in the hearts of all pigs. If pigs wanted to go wild, they would. Wildness was an excess that spilled over human attempts to tame and master it. However, recognition of the limits of their control did not prevent humans from making an attempt to establish tentative relationships of trust and even friendship. Prema had no illusions that she could change her pig's ways by force or by persuasion. Her offer of a potato might be read as a bribe, something to induce the pig to return home when he grew weary of life in the forest. But perhaps it is better to think of the potato as Prema intended it, as an offer to maintain a difficult and fragile *relationship* (*rishta*) that recognized and even respected the exigencies of difference. It is to another such relationship that emerges across difference—that between women and bears who are said to have sex with them—that I now turn.

The Bear Who Loved a Woman
The Intersection of Queer Desires

It was the time of year when *mandua*, beloved food of bears, was not yet quite ripe. The clawlike spikes at the top of the stem that give the grain its name—finger millet—were studded with bright-green seeds that resembled clusters of rice pearls. I had spent the afternoon with a group of women who would meet every week in Chirag's courtyard and knit toys, scarves, and sweaters for its store. It was one of those quintessential late summer days in the mountains when the clarity of light was just astounding; a kiss of the sun's rays on the metallic needles the women held deftly between their fingers made shadows dance across the slate floor. The sun had dipped low in the sky by the time the women finally began to stuff their half-finished projects into faded blue plastic bags. I was staying with Rupa *di*, one of the women, that night, and was anxious to leave since we had a long walk through the forest ahead of us. Mohini, one of the younger women in the group, who loved ribald jokes and shared them with an exaggerated vulgarity that reduced everyone else to helpless giggles, walked out of the gate with us. "Watch out for the *bhalu* (bear)," she said to me, patting my cheek gently. Her brown eyes were alight with wicked laughter. "I'm sure he would love a city girl to do his work (*kaam*) with," she said. A broad wink and then she was gone, walking hurriedly back to her own house. I assumed Mohini was referring to the Asiatic black bear who had recently attacked a woman cutting grass in Kasidunga, a populated area that abutted the IVRI forest, but I was utterly confused by the reference to "work" and her meaningful wink. I looked to Rupa *di* for clarity. Her brow was furrowed. Rupa *di* was the epitome of dignity and generally disapproved of Mohini's crude jokes even though she would often laugh at them alongside the rest of us. "What did she mean?" I asked.

"Nothing. She's bullshitting [*bak rahi hai*]."

But I was persistent, and asked again why I, in particular, needed to be watchful of the bear.

Even though we were alone with not a soul in sight, Rupa *di* looked suspiciously at the dark forest of tall rhododendrons, oaks, and cedars encircling us and then lowered her voice before responding. "It's like this," she whispered, "bears do to women what men do to women."

I was intrigued by this statement, but couldn't grasp its importance. When I told Rupa *di* that I didn't understand, she looked exasperatedly at me, impatience etched into her features. "Why don't you understand, Radhika?" she said, handing me her bag of knitting as she shook out her thick brown hair and twisted it into a tight knot. "*Woh unke saath* sex *karte hain* [they (bears) have sex with them (women)]."

"What? How?" I was stunned by this revelation, and my mind was abuzz with questions.

"*Bhalu aadmi ka rup hote hain*" (bears are a form of man), Rupa *di* said, now with a sly smile on her face. She had clearly forgotten about lowering her voice. "If they find a woman, they forget everything else and enjoy themselves. Sometimes a woman dies when a bear has sex with her. But the bear doesn't stop. He has sex even with the corpse."

Rupa *di*'s tone was factual, as if she were describing a recently transpired event, but I could not figure out what to make of what she had just told me. I asked whether this was a *kahani* (tale). She looked offended. "No, it's the truth," she said. "I've heard of many women to whom this happened. Ask anybody." She paused and then clarified her statement. "Ask any woman."

So I did. A week or so later, I was sitting across from one of my closest friends, Gita, helping her to prepare dinner. She had found some amaranth leaves near the cowshed that afternoon and was washing them carefully in a faded plastic bucket with jagged edges. I had been tasked with cleaning black soybeans that she had dumped from a jar onto a plate. As I sat on a damp gunny sack and picked out little stones from among the beans, my mind went back to the *bhatt ki dal* (soybean gravy) I had eaten at Rupa *di*'s a few nights ago, and with that I recalled our conversation. Gita, I decided, was a good person to ask. She was frank about matters related to sex and spared no one her acerbic tongue, not even her husband of little more than a year. "Listen Gita, I heard something about bears," I began clumsily.

"Yes?" Her tone was distant.

"About what they do to women." She looked up at me now, smiling broadly. "Heard or found out for yourself?"

"Don't be ridiculous," I said. "Rupa *di* told me."

"Did she meet a *bhalu*?"

Our laughter was loud enough that we both looked over our shoulder to make sure her stern father-in-law had not heard us.

"Is it true?" I asked.

"About bears? Yes, it is true."

"What do they do?"

She stopped chopping the greens and put the knife, now stained a dark, vegetal color, down. I, too, put the plate of soybeans on the ground in front of me. Gita was a gifted raconteur, and I wanted to write down what she said.

It so happened that one of my neighbors was married into a village near Kanda in Bageshwar. Once, when she came back home . . . I was very young then . . . she told me about what happened to a woman who lived in a village near hers. One night, this woman was washing the dinner dishes outside the kitchen [near the back of the house]. Suddenly a *bhalu* came and took her away. He kept her in his cave and rolled a big stone over the entrance when he went out so that she couldn't run away. And then he had sexual relations [*sharirik sambandh*] with her. Lying down [*leta ke*]. And then, when he felt loving, he would lick the soles of her feet . . . Men go to sleep when they are finished. But this bear would keep loving her [*pyar karta rehta tha*]. And he would feed her the same things that he ate—*hisalu, bhamora, makka.* It went on like this for months. But then, one day, there was a man whose cows had strayed farther into the forest than they usually did . . . he found her when the bear had gone to hunt. They say that she looked like a bear herself—matted hair and dirt all over her body. They took her to a hospital in Almora. And you know what? She was pregnant!! Two months later, she gave birth to three babies—just like little bears . . . But you know the most interesting thing? My neighbor told me that the woman was sad when they took her away from the bear. She had begun to love him.

"Really?" I asked.

"Yes, they say her husband wasn't nice. Maybe the bear was a better husband."

At this she started to guffaw, and her laughter was so infectious that I couldn't help but join in. I didn't hear her father-in-law come up behind us, and jumped when he asked me why we were laughing so loudly. Heeding Gita's look of warning, I said that we were talking about something funny the cows had done that morning. He looked unconvinced but returned to his room after telling Gita to get on with dinner. "If only a bear would take him and shut him up in a cave," Gita said, with feeling, bringing on another bout of laughter.

These two initial conversations and then subsequent ones with many other women and some men left me with a series of questions about this genre that "belonged" to women across castes (Flueckiger 1996). Some of these queries are inspired by a body of feminist scholarship on how gender comes to be instantiated as a critical genre in the folklore in South Asia, a genre that is not only a "response" but also an "alternate" to "official" discourse and norms (Ramanujan 1991b).[1] Following these insights, I explore how these tellings transgress dominant gender roles and ideologies that treat women's sexuality as a problem even if this transgression is only momentary. On a related note, mindful of the caution that these songs and stories do not just *reveal* the existing nature of lives but are "lived practices" that *make* relationships, I am moved to ask not only what this genre indexes about women's sexual desires and their ideas about love and marriage but also how it *creates* and *shapes* them. Recognizing these stories as attempts to "reformulate the world" (Blackburn and Ramanujan 1986), I ask what the new worlds imagined in these tellings look like and how they address the insufficiency of this existing world. In particular, fueled by the queerness of these tales, I draw on recent work in queer theory to ask what horizons of possibility were sketched through the telling of these stories and how those futures full of potential related to the present that these women inhabited.

These were women's stories, but they were also stories about bestiality that called the boundary between human and animal into question. As such, they raised compelling questions not just about women's lives and desires but also about interspecies relationships. What about bears made them the choice for partner in this carnal interspecies encounter? How did the lustful bears of these stories relate to "real" bears, the Himalayan black bear that roamed the forests of this region? How did bears come to be related to women as husbands? What, in particular, was it about embodied interspecies encounters between women and bears that allowed for the queer boundary crossing— between human and animal, between mundane and extraordinary, between drudgery and love—that is at the heart of this genre?

In working through the answers to these questions over the rest of this chapter, I follow Ann Gold (1997, 108) who asserts that women's stories and the demands and desires articulated through them should not be falsely separated from "the social universe that produces them year-round and also values them." The critical analytical and empirical frameworks that I offer the reader draw on my commitment to situating these tellings firmly in the particular social world in which they came to life. I take these tellings as attempts to imagine and enact alternate ways of dwelling and connecting in a world where one exists only in relation to a variety of human and nonhuman others. These tellings not only have implications for understanding how women pose themselves as sexual subjects in a social world that frowns on feminine expressions of desire but also point to the ways in which feminist perspectives can be enriched by moving beyond the limitations of an anthropocentric approach to understanding women's consciousness and considering how human subjectivity, in general, is crucially formed in relation to the nonhuman world.

Bhalu ki baat

Bears are favored creatures of cultural lore and practice. In legends, songs, and ceremonies of the Northern Hemisphere, they appear most often as sacred, powerful, mysterious, and sagacious characters.[2] For instance, Native American legends about bears abound and represent them as spiritual guardians of the forest who are able to speak with humans and establish kinship with them. In comparison, bears do not command as rich a folklore in India. There are scattered references to bears in folktales around the country, although they figure in these mostly as incidental characters.[3] Where they appear as major characters, such as in one of the Jataka tales, bears are characterized as greedy animals who suffer as a result of being unable to curb their appetites. There is also, of course, the famous anthropomorphized army of bears led by Jambavan that joins Hanuman to help Rama defeat the king of Lanka, Ravana, in the Ramayana. On the whole, though, especially when compared with other animals such as monkeys or snakes, bears make relatively rare appearances in South Asian lore. The genre of stories I describe in this chapter are, therefore, quite unusual in their selection of animal protagonist. Given their close resemblance to humans, however, it is perhaps unsurprising that bears are the natural choice for stories about animals who have sex with women. As Wendy Doniger (1999, 177) points out, it is our feet, the "sign of our human condition." that are thought to separate us from animals. However, bears' occasional bipedalism transgresses that boundary of the foot in evocative ways.

Stories about women being abducted by and having sex with bears or anthropoid creatures bearing close resemblance to bears have long circulated across the Himalayan region. In the 1889 *shikar* narrative *Hindu-Koh: Wandering and Wild Sport on and beyond the Himalayas*, Donald Macintyre, a colonial hunter, noted that "fables innumerable about bears" are common among *paharis* in Kumaon. He writes that these tales are "often amusing from their utter absurdity." As an example of one such fable, he says that "the occasional abduction of women from the villages by bears is firmly believed in." In Nepal and Tibet, the wild man figure of the Yeti, who was believed to abduct women and have sex with them, has become an icon of popular culture since his mysterious traces in these mountains first fascinated generations of colonial officials and European travelers. Some have suggested that the Yeti is simply a representation of people's experience with bears, either the Asiatic black bear or the brown bear (sometimes called *lal bhalu*, or red bear).[4] In an interesting gender inversion of both the Kumaoni genre and the Yeti stories, Tamang villagers in Nepal believe that the *nyalmo*, a female anthropoid creature, abducts men and holds them captive, even bearing their children (Campbell 2013).

Tales of bestiality with bears also abound in medieval Europe. Bears were thought to be lustful creatures, obsessed with seeking carnal pleasure. Bieder (2005, 80) notes that Saint Damian reminded the flock of the bear's carnal nature when he told them that Pope Benedict had been turned into a bear in the afterlife as a consequence of his lustfulness on earth. Indeed, it was the lusty nature of bears that was foregrounded in reports of their abduction of young women. Edward Topsell, in his famous The *Historie of Foure-Footed Beastes* ([1658] 1967, 27), noted that bears were of a "most venereous and lustful disposition." He shared a story of a "young maid" in Savoy who was carried by a bear into his den, "where in venereous manner he had the carnal use of her body and while he kept her in his den, he daily went forth and brought her home the best Apples and other fruits he could get . . . always when he went to forrage, he rouled a huge great stone upon the mouth of his den." Topsell also described the copulation behavior of bears, saying that "the manner of their copulation is like to a mans, the male moving himself upon the belly of the female, which lyeth on the earth flat upon the back."

Topsell's account, in particular, bears striking similarity to the tellings that I heard in contemporary Kumaon—there is the stone that the bear rolls in front of the cave, the little detail about bears copulating like men in the missionary position, the gifts of food, and the general characterization of the bear as a lustful creature obsessed with carnal fulfillment. But what does one call these accounts? Initially I persisted in called them folktales, even after Bimla *di*'s pointed rebuke that she was recounting to me a "true" event. But my characterization of this genre as one of folktales was resisted by all the women who told me of these "events"; this was *sach* (true), they insisted, not an illocutionary *kahani* (tales), a claim they backed up by specifying that these were events that had occurred in the recent past in nearby villages.

Further, when I asked whether these stories were another form of the popular animal-husband or animal-lover tales in Indian folklore, my interlocutors were quick to criticize me for my slow wittedness.[5] Bears, I was told rather acerbically, did not turn into men even though they behaved like them. These were stories about real *bhalus*, the bears who roamed these jungles, not about men who went about in animal form during the day and took their real form at night, and certainly not about anthropoid creatures like the nyalmo or the Yeti. Duly chastised, I finally began referring to this genre as *baat* (happening, matter, thing, talk), a more neutral term than *kahani* (tale), in everyday conversations. In this chapter, too, I resist using the terms *folklore* or *folktale* in deference to my interlocutors' thoughts on the matter. Instead, I use the term *genre, genre of stories*, or even *story* for convenience.[6] However, despite avoiding the term *folklore* for this genre, I do borrow tools and terms from scholars of folklore to analyze it. Thus, for instance, I follow A. K. Ramanujan (1991a) in describing each story I recount in the pages that follow as an "individual telling" as opposed to a variant that diverges from an authentic, original story. And while these stories were not performed during ritual or otherwise auspicious events, I draw inspiration from scholars of South Asian folklore in paying particular attention to the contexts in which they were told and to whom in order to understand what they might mean and what work they do.

The Sweet and Sour of Marital Life:
Unpacking Sexual Violence in *Bhalu ki baat*

When I first heard stories in this genre, I could not help but think that what they were describing was sexual assault—a woman is abducted and must then have sex with a bear. Given my own visceral reaction, I was surprised that, for the most part, the women who told me this story did not represent it explicitly as an act of sexual violence. They used the neutral term *le gaya* (took away) instead of *apharan* or *kidnap*, words that they used in other contexts. But, more strikingly, most women did not use the words *rape* or *balatkar* in telling these stories. This was certainly not because they did not understand the concept of rape; on the contrary, especially with the intense focus on violence against women in newspapers and on television in the past few years, these women were fluent in the language of sexual assault. They often talked explicitly about women getting raped without employing any euphemisms to describe this gendered violence.

Despite this, it was seldom that I heard this encounter between the bear and the woman he "took" described as rape. The word *rape* in the context of the bear's story came up one day when I was invited in for a cup of tea by an eighty-year-old grandmother whose house I had passed during a walk through the village. I walked down the narrow mud path to their house, admiring the masses of chilies drying on the slate roof, and was met in the *aangan* (courtyard) by the grandmother, who whispered to me that I should teach her granddaughter "some good things." I knew what she meant, of course. The whole village was talking about how the granddaughter had recently been spotted in the forest holding hands with and kissing a boy from another caste who worked with her at a rural call center; the girl's family was now forcing her to quit her job and get married. As we drank tea, the grandmother told me about bears, at my request, and ended her story about the bear with the ominous warning that young girls who roamed (*ghoomna*) around the forest would get raped. On another occasion, a young white Canadian woman who was volunteering with the local NGO was doing a "homestay" with one of the families I lived with. On one afternoon during her visit, I was discussing the story of the bear with my friend Gita. The Canadian was keen to participate in our conversation and asked what we were talking about. In broken English, Gita explained that we were talking about bears. She thought for a moment about what word to use and finally said "bear raping woman." In Hindi, however, she had said "bears doing to women what husbands do to women," a phrase most others used as well to convey the sexual nature of this encounter. When I asked if she would characterize the bear as a rapist, she was ambivalent. "Bears are lovers/loving (*premi*). Yes . . . maybe rape. But he takes care of the woman, licks her feet. She starts to like him, too."

The uncertainty of this explanation was mirrored in the broad comparison that women drew between bears and husbands. Bears' intentions were often described as *kharab* (bad) or *khatarnak* (dangerous), semantically fluid terms that I had heard some women use to portray their husbands' excess of sexual desire as well. Perhaps this comparison between husbands and bears was an oblique way of referring to the ambiguous nature of the violence that inflected this encounter. In a few cases all women agreed that there was nothing ambiguous about the terrible domestic and, more specifically, sexual violence that some women were subjected to by their husbands. But more often, husbands were spoken of as mercurial, complex creatures who were loving and indulgent in private but unyielding and unsympathetic in public. Even the best of husbands could participate in the public disciplining of wives, most often at the behest of their parents. This disciplining could take the form of a scolding (*daant*) or the rare beating (*maar*) for some unforgivable infraction.

Manju, a newly married woman who had been slapped by her husband, Dinesh, for lingering over a conversation with the village rake, told me and a group of other young women that her husband had apologized to her that night. She said her mother-in-law had forced Dinesh to slap her and that he was afraid of standing up to his mother because it would only create more difficulty for his young wife.[7] "He said," she continued, "that his mother and other women in the village already think that he is under my sway. She would have made our life hell if he hadn't slapped me at her behest." Even though Manju agreed that her mother-in-law's sermons would have been unbearable if her husband hadn't acted in the way he did, she was still angry with him for his unwillingness to stand up to his mother. "I know men have to do this to show others sometimes or else they'll become a joke," she said to us, "but couldn't he have given me a gentler slap?" Her anger had been somewhat assuaged by his efforts to show his remorse—he had bought her a new *saree* and a pair of silver anklets and had also expressed his penitence by kissing her hands and feet repeatedly. One of the other young women, who had been married for a couple of years, told her to forgive him—"aadmiyon ki jaat hi aise hoti hai," she said, "kabhi khatti, kabhi meethi" (men are just of that type, sometimes sour, sometimes sweet).

Sometimes sour, sometimes sweet. The phrase glossed over, I thought, the very real violence that was visited on Manju and others like her. But, it was also a reminder that many women saw everyday and, sometimes, exceptional acts of violence as constituting one element of a complicated relationship with their husbands. While Manju was still smarting over the slap, she also portrayed it as essential if Dinesh was to avoid becoming the laughingstock of the village. Upon listening to her justification of Dinesh's actions, I was reminded of Laura Ring's (2006, 113) observation that men's anger can often be viewed not just as "an inevitable facet of masculinity, but as a *necessary*, even foundational one." It is condemned and lamented but also cultivated as a crucial part of masculinity. These moments when a violent and toxic masculinity reared its ugly head were the *khatta* side of matrimony. But it was only if you had a taste of the *khatta*, many women tried to convince me, that you could savor the *meetha*. This is not to say, however, that women always accepted the ratio of *khatta* to *meetha* that their husbands doled out; as I discuss in the next section, there were constant attempts to negotiate this balance in everyday life. As I struggled to understand the meaning of the story of the bear who did to women what husbands did, I thought perhaps it was a story that indexed the *khatta-meetha* nature of intimate relationships. All women would agree that there is an unmistakably violent dimension to the tale. But, in the story, the bear is also gentle and loving. Perhaps the more important question than what this *khatta-meetha* genre of stories meant was what work it did in the *khatta-meetha* of everyday life. It is to this question of *doing* that I now turn our attention.

Sexual Appetite and Gendered Critique

It was a hot evening in June. The air was heavy with humidity even after the sun had sunk low in the sky. The village of Pokhri was in the throes of an acute water crisis, and I was accompanying my friends Leela and Prabha to the village hand pump, the only source of water near their home. There were several other people there already, so we retreated to the welcome shade of an old horse-chestnut tree to wait for our turn. I was drenched in sweat and could no longer stand to soak it up with my already sodden *dupatta*. "And what's new [*nai-taji*; literally, what's new and fresh]?" I asked, more to distract myself from the sticky heat than from any real desire for gossip. Leela, who was married to Prabha's cousin, started to tell us of a fight that she and her husband Mohan had had a few nights ago. It had begun when Leela had caressed his body and pressed herself against him. Mohan responded by inquiring when she had become so forward and with whom she was learning these new tricks. He then told her that he was too tired to have sex with her that night, and that she would have to wait for him to recover his energy. This annoyed Leela immensely, and she retaliated by saying that she didn't have another lover yet but would have to start looking for one if things stayed the way they were. Mohan became angry and twisted Leela's arm, telling her to watch herself and remember that he was her *malik* (a term that is used to signify both owner and husband in this region). But Leela would not back down. She wrenched her arm free and told Mohan that a bear would make a better lover (*premi*) than him. "And then I said 'at least the bear will not be tired all the time. He'll be ready whenever I want him.'" I asked how her husband had responded, fearful that things might have become violent. But Leela assuaged my concerns and said her husband had first started to laugh and then finally said that she didn't need to find a bear because he was ready. "And then . . ." Her voice trailed off, and she gave us a sassy wink while miming a kiss. "Chhee chhee. Bhabhi yaar," yelled Prabha blushing furiously, "The things you say. Others might hear you." "So what? What if your *bhabhi* had to run away with a bear? What would you say to others then?" I couldn't help but laugh at the satisfied expression on Leela's face.

What do we make of the invocation of the bear in this moment of marital friction? One could read Leela's "sexually irreverent discourse" as an act of subversion against a powerful gender ideology that treats women's sexuality as a risk to society at large, as something to be reviled, an ideology that we also saw at work in the response to Manju's overlong conversation with the village lothario (Abu-Lughod 1990, 45).[8] In this particular case, though, it was clear that Mohan's quick, initial response to Leela's overtures—the accusation that she was having an affair with another man—sprung from his fear that she had an excess of sexual desire that was not exhausted by the scope of his own desire. He saw this desire as errant and outside the bounds of the "sexual economy" of marriage (Ramberg 2014)—one where sexual exchange is driven by the husband's desire—and it was this that led him to punish her by reminding her that they would have sex only when he was ready for it. Even though theirs was a "love marriage," Leela's unabashed sexuality was discomfiting to his sense of control and, indeed, ownership of her.

The image of the sexually ravenous wife as a threat to the family and society has a long and complicated history in South Asia. As Raheja and Gold (1994, 10) note, "both Sanskrit texts and vernacular oral traditions contain positive images of women as mother and as ritual partner and exhibit disdain and reproach for women as wives who are seen as sexually treacherous, sexually voracious, and polluting by virtue of their association with menstruation and child-bearing." Uma Chakravarti (1993) makes the important point that the reason sexual purity became such a foundational element of Brahminical patriarchy, as outlined in normative texts such as the Bhagavad Gita, was because the maintenance of sexual control over wives was crucial not just to ensure patrilineal succession but also to preserve caste purity.[9]

Even though the supposedly natural proclivity of wives for sexual perfidy was thus a long-standing concern in South Asia, feminist scholars have illuminated how establishing the chastity of wives became a particularly important political, social, and legal project in the colonial period. Writing about the reform efforts of nineteenth-century Bengali nationalists, Tanika Sarkar (2001) points out that these overwhelmingly upper-caste men believed that "successful governance of the household" would be a first step toward establishing a claim to political power and, especially, to constructing a Hindu nation. The first step in this nation-building project was securing the chastity of the wife, a project that, as Lucinda Ramberg (2014, 159) has so deftly shown, entailed opposing her sexuality, a gift given to her husband, to that of the prostitute, a figure who was now "defined by her perverse sexuality" which was immoral, illegitimate, and dangerous. The virtuous wife thus came to be emblematic of the idealized Hindu nation, her monogamy its very condition of possibility (Sarkar 2001, 167). Sexual control, however, was not a feature of upper-caste reform efforts alone. As Anupama Rao (2009, 53) observes, even "caste radicals [who] were preoccupied with challenging caste ideology by rethinking marriage and sexuality . . . were by no means immune to the extension of novel patriarchal practices into their own households."

The figure of the virtuous wife, which took shape at the juncture of colonial anxieties about sex and moral degradation and nationalist and reformist idealizations of chaste women as the bedrock of community and nation, was held up as the cornerstone of the family in contemporary Uttarakhand. This was true across castes; indeed, a number of lower-caste families expressed immense pride in the fact that their daughters-in-law wore a *ghoonghat* (veil) in public. Good wives were supposed to bear children, preferably sons, but simultaneously express sexual restraint.[10] In particular, they were expected to cultivate and embody the feminine virtue of *laaj*, which can variously be translated as "shame," "modesty," or "honor."[11] Young wives were expected to refer to their husbands only in the third person or as the father of their eldest child, never by name; to cover their heads in the presence of their *jeths* (husband's older brother); and to avoid any prolonged interaction with men from outside the family, among a host of other prescriptions. Any lack of appropriate display of *laaj* was thought to signal that a woman was slipping off the yoke of her family's control. Indeed, I was sometimes told that I displayed a lack of *laaj* when I referred to my partner by name instead of in the third person; however, I was easily forgiven for my infractions because I had grown up in the city and now lived in America and clearly did not know better.

Despite the social force of these expectations and the generalized acceptance of the necessary anger of husbands, the women I knew certainly did not resemble the stereotype of submissive and repressed subjects under the thumb of their fathers and husbands, an image that is still popular in orientalist, colonial, and even feminist scholarship.[12] Even as many women internalized and deployed these ideologies of control against one another in particular situations, they were also constantly negotiating and critiquing them in other social contexts. Uma Chakravarti's (1993, 580) observation about the urgency displayed in Brahminical texts on the question of controlling women's sexuality is apposite here. She writes that "the detailing of norms for women in the Brahminical texts are a powerful admission of the power of non-conformist women, or all women who have the power to non-confirm." Similarly, the ubiquitous reiteration of discourses about control over the sexuality of women in Kumaon bears testament to the fact that women regularly and joyously flouted such conventions and called them into question. Put another way, it was clear that women were quick to protest when an excess of *khatta* was injected into their lives.

This contestation played out not just in feminine, intimate spaces but also in more public, mixed-gender arenas. On one afternoon, I watched as a young woman, Kamla, whose mother-in-law had berated her for sleeping next to her husband in the afternoon and "disturbing him," took offence at the accusation that she had displayed a lack of *laaj* in acting as she did. "What is the meaning of this, Ijja?" she shouted, loud enough to attract the attention of the neighbors, her father-in-law's older brother's family, whom I was visiting for a glass of tea. She was standing in the *aangan* of the house, a space that was visible from the main village thoroughfare that lay right above their house. Her mother-in-law glanced up at the road, visibly embarrassed that the people who were walking on the road could hear and see her daughter-in-law. "I'm not dying to sleep with him [her husband]," Kamla continued angrily, drawing a shocked gasp from her mother-in-law. "Give me another room and I'll sleep there happily without him. I'm the one who has worked hard all day. Why don't you ask him to sleep elsewhere?" Her mother-in-law was about to respond, but Kamla didn't let her get a word in. "Don't you dare say this rubbish [*bakwaas*] to me in the future. Remember I used to work in an NGO. I won't tolerate this talk. False *laaj*, false *sharam*. I'm treated worse than a buffalo in this house. Don't you have any *sharam*?" Her husband, who had been watching from the sidelines, admonished his mother for treating Kamla badly and threatened to move elsewhere if she kept harassing his wife.

By the next morning, everyone in the village had heard about Kamla's outburst. Many younger women and some older ones, too, thought that she had done the right thing. "It's a different time," one older woman told me. "These are *modern* girls. They will turn around and give you an earful. You just should not say this kind of thing to them." I heard similar sentiments on multiple occasions. Many people associated these challenges to the dominant gender ideology with "modernity." "This [attempts to control women's sexuality] might have happened in the past, but it has no place in today's world" was a familiar refrain of young, educated women in particular. It is true that these women's self-imagining as "modern" subjects was, as they saw it, a crucial element in their resistance to what they considered antiquated forms of "tradition." For them, the project of modernity entailed representing themselves as fundamentally different from the generations of women who had preceded them, as more aware of the unfair nature of patriarchy and less willing to tolerate it. However, I want to resist the idea that challenges to this patriarchal social system were new and borne out of increased literacy, mobility, or exposure to "modern" discourses of women's empowerment that circulated in the region through the efforts of local NGOs. For one, as feminist scholars have pointed out, the representation of sexual violence as "backward" and women's empowerment as "modern" is a framing that replicates certain colonial stereotypes and does not allow us to get to the roots of patriarchal violence and power.[13] Moreover, even older women, now represented as the bearers of tradition, had struggled under the weight of these forms of control and had resisted them in their own, sometimes similar, ways despite never having imagined themselves as "modern" in doing so.

Let us return, then, to the genre of stories about bears and women becoming sexually intimate and the question of what work they do in this milieu where women's desire constitutes a problem. Given the weight of normative gender expectations, it would certainly be appropriate to read these stories as narratives of critique that subvert the denial of legitimacy to female sexuality. Scholars of women's folklore and songs in South Asia have pointed out that these expressive traditions provide women with the opportunity to grapple with the social worlds they inhabit, comment on and critique quotidian relationships, provide moral instruction, contest their marginality, and shape their subjectivity. In his classic work on folktales, A. K. Ramanujan (1991b) argues that women's tales constitute "counter-systems [that provide] an alternate way of looking at things." "Genders," he argues, are "genres." This genre, too, was very much a woman's genre, told by women and taking on the female point of view. Much like women's songs in the Kangra Valley that Kirin Narayan (1997) analyzes, these were stories that centrally engaged women's experiences and desires and certainly did not speak for Kumaoni "culture as a whole."

But the fact that these were female-authored-and-narrated stories did not mean that they circulated only in a discrete feminine realm. Men were not unaware of this genre or the subversive nature of its content. These stories could be and often were directed at them, as Leela had done with her husband, to shame them for their lack of sexual desire or their attempts to sublimate female desire to their own. In these tellings, men were confronted with women whose sexuality blossomed through an intimate encounter with a responsive partner. Women spoke with relish of the sexual encounter, sometimes within earshot of or directly before their husbands. Neetu, a thirty-year-old, told me how the "work" continued "dhap dhap dhap [*mimicking the sounds of sex*] from both sides all night long." Her eyes lit up with wicked pleasure as she smacked her palms together in time with the words. Her husband, who was in the next room, walked in with an embarrassed look on his face and told his wife to hurry up with cleaning the stove and get to the fields. She was unfazed and told me, without lowering her voice, that her husband's nightly bouts of alcohol consumption meant that he couldn't do the "work." "Doesn't do it himself and won't let others talk about it," she sniffed, putting yet another kettle of tea on the stove and making herself comfortable on the gunny sack that lined the mud floor of the kitchen. "And then people ask me why we don't have a child yet?" Neetu's allusions to the woman's enjoyment of the sexual encounter with the bear affirmed and celebrated the bear's sexual prowess while simultaneously commenting on her own husband's lack of sexual capability. The fact that this union between the woman and the bear ended in reproductive success—with the woman eventually spawning three children, even though they looked like bears and had bearlike tendencies—made the comparison even more uncomfortable for Neetu's husband. While the social stigma of not having produced children was borne almost entirely by women, Neetu was able to redirect some of the shame and responsibility to her husband through her telling of a story that celebrated both women's sexual desire and fertility and linked them to one another.

I witnessed the subversive power of these tellings on another occasion, which I will describe in some detail for two reasons: first, it was a telling that ended differently from others I had heard; and, second, it was the first and only time that a man would speak to me directly about his views on this genre. I was spending a couple of days at the house of a family with whom I had stayed several times before. Sitting in the kitchen with my friend Kusum, a mother of five girls, I took out my notebook at her instruction. In her telling, a young woman who faced excruciating physical violence at the hands of her husband and in-laws longed to leave her household. She took to walking in her *aangan* (courtyard) every night, delaying her eventual return to the bedroom she shared with her husband, a space that she associated with nightly beatings. One night, a few years ago, a bear took her from the courtyard. Her family searched high and low for her, but could never find her. Eventually villagers started seeing her in the forests around the village. She looked like a bear herself, with tangled and matted hair, sometimes walking on all fours. If people tried to approach her, she would run away and disappear into the cave she shared with the bear. Kusum's reading of the woman's refusal to return was that she was treated better by the bear than by her husband, and she fell in love with the bear. "Would you have come back?" she asked me. "If I were her, I would have stayed there too. Free. Happy. The bear did not beat her. He kept her well."

About halfway through Kusum's telling, I noticed a shadow outside the low wooden door that led to the kitchen garden. Leaning back slightly to see who had come, I saw that it was Kusum's husband, Inder. When I gestured to Kusum that he had crouched down and was eavesdropping on our conversation, she shrugged and continued telling her story, louder than before. When I came back from a walk with two of their girls later that evening, I found that Inder had ripped out of my notebook the page on which I had written Kusum's telling. "Don't write about this," he admonished me. "It's women's nonsense. It's not true," he spat out. "They are uneducated and can't talk about good things so they talk about this rubbish. Do you think such a thing could have actually happened? This is a dirty story . . . these women have dirty minds. But you should not sully your mind with them." I protested mildly that I didn't care whether the story was true or not; I said that I was interested in collecting all the stories that women in the mountains told. But Inder was insistent. I was angry but decided not to make a scene since I was spending the next few days there. Kusum, however, would not let the incident go even after I told her several times that it didn't matter. The next morning, she returned the torn page from my notebook. "I told you this story," she said to me, "and I'm telling you to write it. These men think they can enslave us (*bandi banana*). But I won't tolerate it."

Perhaps Kusum's telling was directed as much at Inder as at the visiting anthropologist. I had known for some time that she resented her in-laws for their ill treatment of her daughters. Her mother-in-law, on several different occasions, had told me that she thought Kusum and Inder should give the girls up for adoption. There was no point, she once said, in raising five girls who could only be an expense for the family. Now five boys, she continued, would have been a different matter, but Kusum's failure to give birth to any boys had brought "ruin" and "despair" to the household. Kusum told me that she was increasingly frustrated with Inder's silence during these cruel diatribes and sometimes thought of leaving him and returning to her natal family. It was the certain knowledge that they would not welcome her return, either, she once said with immense sadness, that forced her to tolerate this daily abuse. She had told Inder on several occasions that she felt imprisoned in her life. Little wonder then that he felt so threatened by her telling of the story in which life with the bear was framed as freedom from the tyranny of the household. What was so interesting about Kusum's telling was that the woman *chose* to stay with the bear, unlike most other tellings I heard in which the bear rolled a rock in front of the cave's entrance to keep her captive, and where the woman was eventually, albeit reluctantly, rescued by villagers and brought back into the fold of family and community. Even in those tellings, the woman's reunion with her human family was not described as a happy one. Indeed, in Gita's telling and in others like hers, the woman was represented as longing for a return to the bear with whom she shared an inexplicable bond. But what was so radical about Kusum's telling, and the fact that she directed it at her abusive family, was that the bonds of family were completely snapped by the woman who was first taken by the bear. When she saw villagers with whom she had lived and worked, she deliberately avoided them. Even though this meant forsaking human society, she had made it clear to villagers that life with and *as* a bear was preferable to life under the shadow of patriarchal violence.

What Do You Think a Bear's Banana Will Feel Like? The Queer Pleasures of *Bhalu ki baat*

The tellings of this story, as I have argued, often had subversive effects, calling powerful gender ideologies into question and critiquing male control over female sexuality. But enunciating their refusal to comply with patriarchal norms and expectations was not the only intent behind women's desire to share these stories with me and with each other and their male relatives. As I listened to multiple tellings, I recognized that they shared something more in common than just the fact that they deliberately gestured toward the flourishing of women's sexuality in the face of sustained efforts to repress it. These were stories that stoked desire and longing, that were driven by the "pleasure principle" (Ramanujan 1991b). Through these tellings, women constituted themselves not just as resisting subjects but as desiring subjects. As women told or listened to stories of this encounter, they imagined themselves in the place of the woman who had been taken by the bear and fantasized about what sex with the bear would feel like—how many times in the day they would have sex, whether the bear's loving bites and nibbles would hurt, whether the bear's fur would feel rough or silky when they ran their hands through it.

Indeed, I was enthralled by the fundamentally queer nature of the pleasure that women articulated in their imaginings of the infinite horizon of sexual possibility that would open up to them through sex with the bear. Sexual intimacy with a bear was a pleasure that the women who told this story had never experienced in embodied form, but their imagining of the forms that this pleasure would take was unencumbered by the limits of experience. Taken together, then, these stories constituted a queer archive of desire that hovered somewhere between an insufficient present and an unrealizable future; in their temporal fluidity, they illuminated "a horizon imbued with potentiality," a phrase I borrow from Jose Muñoz (2009). The women who constructed this queer archive and imagined themselves into the stories it housed thus "undid" (Edelman 2004) their present and worked toward "enacting new and better pleasures" (Muñoz 2009, 1) to fill a world that was lacking in desire. This was, of course, a radically subversive act, but what I am suggesting is that subversion was not always the intent of telling these stories. While all women, including the older ones, who told me these stories savored and drew out the naked sexuality of the encounter, one young woman in particular forced me to think about the fact that this genre could sometimes be largely about pleasure and sexual curiosity.

"What do you think a bear's banana will feel like? Will it be hairy like the rest of him? What will the hair feel like inside?" The questions spilled out of Asha, a nineteen-year-old woman who was to marry a man from a village about a hundred miles away the next month. I didn't quite know how to respond. This was the first time I had met Asha, who was distantly related to a Pundit family that I knew well. When I joined her and her sister-in-law, Meena, in the kitchen, the conversation had begun innocuously enough. Who I was, why I was there, what I was researching. It was when I mentioned bears that Asha had asked me the questions about their penises. I didn't know how to respond given that I didn't know these women at all. "Shameless girl," said Meena, swatting Asha with her ladle. "What will she think of us? She'll think that you can only talk about things like this." "You don't think that do you, *didi*?" Asha looked at me appealingly. I assured them that I wasn't offended or shocked. On the contrary, I said, I was happy that they were treating me like one of them. Reassured, Asha told me that she didn't care if others thought that she was "shameless" for talking about these matters. "I want to know," she said. "When my mother told me about what bears do to women, my first thought was 'what must it feel like?'" Her sister-in-law told her that she would soon know what "it" felt like. But Asha was not to be deterred. She told us that she wasn't interested in talking about men, that she would soon find out as her *Bhabhi* had said. But what she really wanted to know was what it felt like to do it with a bear. "Will he lie down and do it? Or will he do it from behind?" This time Meena's gasp was genuine, and she told Asha to stop this vulgar talk or she would tell the rest of the family. Later, as Asha walked me up to the road, she asked again if I was offended by her talk. "Don't mind, *di*," she said, slipping her arm through mine. "I just keep thinking about it, that's all. Sometimes I want to ask other women. I keep thinking what if he takes me. Will I enjoy it?"

I was quite taken with Asha's frank and unapologetic curiosity about what the experience of having sex with a bear would feel like. For her, listening to different women's narrations of this tale was a thrill, almost as much of a thrill as it was for me, the anthropologist, to run into new tellings. Asha told me that she had asked almost every woman she knew about these stories. "Everyone says something different," she told me the second time we met, after a *puja* where she had proved to be the most enthusiastic and knowledgeable member of the *bhajan mandli* (devotional singing group). We were sitting on a grassy ridge with her friend, Rekha, watching the sun set. The twisted bands of color that remained in the purple sky after the sun had disappeared behind a row of hills reminded me of an agate. I could hear a dog barking furiously in one of the houses scattered in the valley below us and wondered idly if a leopard had wandered into the village. Asha was tapping my shoulder now, keen to regain my attention. "Some people say that a bear can draw milk from a woman's breasts, even if she isn't pregnant. That's how much love he sucks with." Rekha, who was despondent that her betrothed (her wedding, too, was scheduled for the next month) hadn't called her even once after their engagement, muttered that she was probably better off marrying a bear instead of the boring man she would have to live with for the rest of her life.

Asha was disgusted by Rekha's digression. She informed Rekha that even her husband to be never called her. But that was no reason to not focus on the matter at hand. "I'm talking about bears," she said in a huff, "don't you ever think about it? Why do they want women? What do women have that a female bear [*aurat bhalu*] doesn't? And then I think what do bears have that men don't?" Rekha, despite still being in a funk, was intrigued, and she and Asha spent the next half hour coming up with possible answers to the question. As I listened to them, I was struck by how full of pleasure the conversation was. Through their talk, these young women were constituting themselves as erotic subjects in search of sexual gratification. One could even say that in taking such "pleasure in the confusion of boundaries," they were enacting the "cyborg politics" that Donna Haraway (1991) advocates. It was that evening when the fundamental queerness of these musings, in the sense that they called a pleasure-filled future into being, impressed itself on me. The bear in these stories, I realized, was a messenger of that future saturated with hedonism and experimentation. But why bears? What allowed the ursine residents of the woods that surrounded these villages to act as such powerful auguries? What was it about bears as individuals and as a species that motivated these desires? The question was always at the back of my mind but moved to the front of it on that very evening when Asha and Rekha asked me whether American bears had the same habits as their bears. This question was followed with multiple others about bear biology. I had to admit that I had not the faintest idea, but I promised to find out.

Bear

Of the three species of bears who reside in the Himalayas—the sloth bear, the Himalayan brown bear, and the Asiatic black bear—it is *ursus tibetanus*, the black bear, who is the most common resident of the Himalayan foothills where I conducted fieldwork. Indeed, the Himalayan region, particularly Kashmir, is thought to be home to one of the largest populations of black bear in Asia.[14] Black bears can grow quite big, with some adult males reaching almost 450 pounds. Sexual dimorphism, as in most species of bears, means that females are usually quite a bit smaller. As the name suggests, these bears are outfitted in a glossy black fur coat except for a V-shaped patch of white at the chest. Small ears top a head that is broad, almost round, and encircled by a shock of hair that resembles a Victorian ruff. Their eyes, too, are small and almost disappear in their face. In the winter, their hair grows longer to protect from the cold, giving them a rather fearsome appearance. One villager described a bear he had seen at the end of winter as a *chudail*, a witch who is distinguished by her long, unbound hair.

Black bears are omnivorous and consume a diet rich in plants, nuts, honey, fruits, insects, and carrion. As was the case with wild boar, villagers told me that some foods were irresistible to bears—millets like *mandua*, the ripening scent of which is thought to drive them wild; *bhamora* (*Cornus capitata*), a wild fruit; corn, in search of which they will enter even the heart of a village; and apricots and peaches, which people claim bears will savor slowly. Indeed, the reputation of black bears as connoisseurs of good food precedes them. One colonial sportsman described "Bruin" as a "thorough gourmand [who] shifts his quarters so as to be within reach of the delicacies of the season, whatever they may be."[15]

Not all colonial sportsmen and officials were indulgent of Bruin's gourmet tastes. Most decried black bears as inveterate crop raiders who caused irreparable damage to crops. Along with other "dangerous beasts"—which included tigers, elephants, buffaloes, wolves, and leopards—bears were a target of the colonial state's early vermin eradication programs carried out before the era of wildlife protection described in chapter 5 was ushered in. These programs were carried out primarily with the help of native *shikaris* who, in many cases, were rearmed for the purpose.[16] The eradication of vermin, as K. Sivaramakrishnan (1999) notes in his study of colonial forestry in Bengal, was a cornerstone of colonial "statemaking," an iconic demonstration of its control over the wild beasts who populated its territory. There was also the matter of securing the empire's revenue by eradicating species who encumbered the spread of cultivation and the management of forest plantations.

In 1912, at a time when protection of game was being hailed as of the way of the future for the reasons that chapter 5 details, colonial officials in the United Provinces began to discuss the possibility of reintroducing rewards for killing the "Himalayan black bear."[17] The conservator of forests wrote to the chief secretary of state pointing out that the "Himalayan black bear is responsible for very serious damage to both field crops and forest trees in the hills."[18] The damage, he noted, was especially egregious in the case of deodar and blue pine trees. Both trees, of course, were important for colonial timber operations at the time and would become all the more so in two years, when the First World War would break out. Another official chimed in, calling the black bear a "savage brute" and suggesting a reward of three rupees per bear to ensure a sufficient number were killed.[19] The conservator of forests responded in May that year by reintroducing a reward of three rupees for the "destruction of Himalayan black bears" and a special reward of twenty-five rupees for individual bears found to be causing serious damage to the forests.[20] Donna Haraway's (1989, 10) observation that the primate body "as part of the body of nature, may be read as a map of power" is useful here for understanding these colonial concerns about controlling and, indeed, destroying the ursine body. Colonial power was effected through the hierarchical ordering of traffic between differently raced, sexed, and gendered bodies, both human and nonhuman; in this case, the black bear's body had become a fertile site for the inscription of difference between nature and culture, colonizer and colonized, and civilized and savage.

Forest officials and conservationists continued to support bounties on the Asiatic black bear in the Himalayas even after independence.[21] Even though bears were granted protection under the Wildlife Protection Act of 1971, their numbers have dwindled since colonial times, and they are now listed as endangered. It was certainly rare to see bears in the villages of Kumaon when I did fieldwork; many younger villagers had never seen one in their lives. However, they did remember their grandparents' stories of hunting these animals alongside colonial officials. But even though these creatures were rarely seen, they left tangible traces of their intimate presence—mangled stalks of corn in the middle of a field, flattened to the ground as if the bear had pulled them out and laid down in them; a bear's ground nest in the forest that was warm to the touch; an impress of a paw in a puddle that was frighteningly close to a village home. Perhaps one glimpsed a bear fleeing through a field of tall maize, as my friend Gita did late one evening; she told me to be careful when I walked around in the evenings because this bear would come to their fields every year without fail when the maize ripened. Juno Parreñas (2016) reminds us that such encounters are no less meaningfully embodied or individuated even though they engage bodily traces rather than actual bodies. Bears were dangerous animals that came to be known intimately through such encounters with their traces even if they did not always make themselves visible in the landscape they shared with people.[22]

But perhaps the most visceral sign of bears' proximity were the scars they bestowed on the many people who were mauled by them each year in the state. Every few weeks the newspapers would report one such attack accompanied by a graphic photograph, usually of the victim's unrecognizable face. In 2011, I met a woman in Almora district who had escaped with her life after being mauled by a bear more than fifteen years ago. Her face had archived the encounter for posterity, covered in scars that looked angry even after all that time. She described how the bear had come upon her when she was cutting firewood. At first she tried to hit the bear with a stick, but when that only served to enrage him, she decided to play dead. After a while, the bear left her where she was and disappeared into the forest. "I was lucky he didn't try to carry me away with him," she said, "I was so afraid that was what he was trying to do when he kept trying to lift me."

My conversation with her reminded me that the risk of encountering and being attacked by a bear is gendered. As noted earlier, it is women who are almost entirely responsible for *jangal ka kaam*, forest work, which entails cutting leaves and grass for fodder and wood for fuel. It is while doing this labor that they most often run into wild animals, whether boar, leopards, or bears, encounters that can sometimes end badly. Encounters with bears can be particularly risky given that it is hard to escape them if they charge. Women's experience of animals—both wild and domestic—is thus wrought by the gendered nature of this labor. As I argued in chapter 1, this means that they often relate to animals in distinctive ways and with a different intensity than men. What lends stories about bears having sex with women their edge is the fact that meeting a bear and being mauled is not out of the realm of the possible for most women. As far as village women are concerned, these are real stories given credence by the fact that women put themselves at risk through forms of labor that frequently involve encountering if not actual bears themselves then the material traces of actual bears that signal their constant presence in the landscape.

I was reminded how much women are at risk in these encounters after the first time I came across the material trace of a black bear. It was the month of September, when all foods beloved to bears are ripening or ready to eat. It was still early in the evening, and I was walking home on a forest trail with some people when the person leading our group stopped short and pointed at a fresh pile of bear scat right in the middle of our path. The scat was greenish black and studded with little seeds; it was already covered in flies. We took turns to examine the scat, keeping a keen eye on our surroundings the whole time. The bear who had left this trace became familiar to us through our engagement with his/her "bodily effects"; even though we did not share any direct contact, our band of travelers gained "intimate knowledge" about the bear through his/her material residue (Parreñas 2016). Our encounter with bear shit that evening was enough to send a frisson of fear through our bodies. We huddled together as we debated feverishly about whether it was safe to keep going on this path or whether we should retrace our steps and take another route home. We still had a long way to go through the forest, and nobody wanted an encounter with a bear in the dark, least of all on this narrow path. But turning around would lose us valuable time. Finally, we decided to take turns shouting loudly as we hurried home, just in case the bear was still ahead of us on the path. As we walked, we exchanged stories of bears, our fear slowly turning to excitement and wonder. When we finally reached Pokhri, I said a relieved goodbye to the others and turned on to my short path home. But then, my friend Lata came rushing after me and hurriedly whispered something in my ear that returned my thoughts to the potential danger that the encounter with a bear's fresh shit held. "Thank god you and I are safe, *di*," Lata said. "But I'm scared for the women who will take that path home on their way back from cutting grass. I hope he doesn't take them. I'll think of only this all night." I was now worried myself, but assuaged Lata's fears by insisting that the women would be sure to spot the scat and be on their guard.

"It's a good thing you and Lata didn't actually see the bear," echoed my friend Gita at dinner that night, little more than a year after I had first listened to her telling of the story. "I'm sure those women are safe. Otherwise we would have heard . . . If a bear attacks you, then there's death written in your future. Or . . ." she trailed off, "something else." She started to laugh. Luckily for me, she went on, bears preferred women to girls. People in the mountains often mistook me for a *ladki*, unmarried girl, even though I was technically an *aurat*, married woman; indeed, the fact that I did not wear any of the markers of a married woman—*sindoor* (vermillion) in my hair, a *mangalsutra*, or *bicchhis* (toe rings)—made it difficult for people to believe that I was an *aurat*. As far as Gita was concerned, even a bear would not be able to tell that I was an *aurat*. That night, I decided to ask Gita if bears were the only animals who sought out women for sex. I had heard some women voice similar fears about monkeys, wondering whether they, too, "did things like bears." However, no woman I talked to except one was certain that monkeys had sex with women (Govindrajan 2015c); she had claimed to me and a group of other women that a woman in a nearby village had been raped by a monkey and given birth to half-human–half-monkey children. The news was received with a mixture of shock and satisfaction, but more than a few women doubted its veracity even as they relished the possibility of it being true. Most often, the women who wondered about monkeys claimed that they had heard rumors of monkeys "raping" women in cities and asked me to confirm whether this was true. The nature of these rumors, thus, was distinct from *bhalu ki baat*, which was, as I have described earlier, spoken of as a *truth* experienced by village women. These women may not have been surprised to learn that monkeys behaved in the same way, but they did not know this with the same certainty that they did the predilection of bears. With these thoughts in my mind, I asked Gita with exaggerated naïveté whether bears were the only animals who did this "work." "Our grandmothers used to say that monkeys were wicked animals too," she responded, as if echoing my thoughts. "But I've heard only of bears doing this kind of work. It's because they are like human beings." She continued, listing the reasons why she thought bears are like humans. "They stand on two feet. No other animal can do that. And they eat with their hands, just like us. In fact, don't you think they look like us, even more than monkeys? That's why they look for women all the time. There's not that kind of fear with monkeys."

After a few minutes, she asked whether I had ever seen bears mating. I said I had not, and she said that she hadn't either. But her grandmother, she remembered, had once told her about coming across bears mating in the forest. "She said that they were doing it for an hour, she saw the whole thing. The sounds! She told me she felt embarrassed (*sharam*) watching them. They were there the next day as well when she went to the same area to cut wood. She said bears don't get tired. They keep doing it, keep doing it, keep doing it." Gita's *jethani* was there that night, and she added her voice to Gita's. "That's what we heard as children too. That bears are in season [a euphemism for mating] for a long time. Just think, what if they were to do that to one of us. I would die from exhaustion if he was on top of me for three days. My great-grandmother even said," and here she lowered her voice, "the bear's that [referring to his penis] is very long." Gita was fascinated by this piece of information. "But how did she know?" she demanded. "Someone in her *mait* had a pet bear," her *jaithani* recalled as she poured some milk into a steel plate for the dog's dinner. "He was very loving, she said. *Hamesha dadi-moti karta tha* [he used to hug people all the time]. One day my grandmother saw it. It was this big [*stretching her hands*]. She called all her friends, and they just died laughing."

When I finally started reading biologists' accounts of the mating habits of bears, I realized that Gita and her sister-in-law had actually learned quite a bit about bear behavior from their elders. It turned out that stories about virile bears who wouldn't tire of having sex actually resonated with observations of the mating behavior of bears. Bears usually copulate for long periods, sometimes hours. The reason for this, as biologists have noted, is that male bears have a long and thin retractable bone called the baculum that during mating enters a sheath in the penis and makes it stiff, which is thought to enable the animal to mate for longer and allow it to deposit sperm directly in the reproductive tract of the female. This is, as Robert Bieder (2005, 23) notes, "characteristic of species where ovulation is induced through intercourse. The baculum is thought to provide the longer and more intense stimulation needed for ovulation."

The genre of stories about bears having sex with women, then, drew on a body of local knowledge about real bears. While villagers might not have known about the baculum, they did know that bear sex was long and intense, unlike other species they were familiar with. These stories were not fantasy, at least not when it came to what they described about the behavior of bears. Instead, the picture of bears that emerged in them—occasionally aggressive, inclined to attack women, and sexually active—was sketched from people's everyday experience of living in proximity to these animals. Even though bears were not as ubiquitous as monkeys, wild boar, deer, jungle cats, or even leopards, theirs was a tangible material and symbolic presence in the landscape that could not be ignored. Villagers and forest officials both swore that the forest was thick with bears; that their trace, if one looked carefully, was everywhere. There were always moments in one's everyday life that were ripe with the possibility of encountering them even if they did not come to fruition. Whether or not they had ever seen them, people constantly shared knowledge about them, not just to avoid potentially fatal encounters but also because they felt a genuine curiosity about their lives. Perhaps it was this widespread curiosity about actual bears—mysterious and powerful figures—that made the stories about them having sex with women so gripping, whether one celebrated those stories like so many women or denounced them like so many men.

Desire as the Root of a New World

What insights might this genre yield for feminist scholarship on the worldmaking power of women's genres in general? In this chapter, I have suggested that talk about bears who have sex with women has powerful effects on the gendered hierarchies that circumscribed the everyday lives of women in the mountains. *Bhalu ki baat* (the matter of the bear/talk of the bear) allowed women to mount a radical critique of rigid notions of sexual purity and control that portray sexually curious and voracious women as a stigma to their family and community. Through telling these stories about a fundamentally transgressive intimacy between women and bears, these women reimagine themselves as subjects who are constituted through their sexual desire. Women and bears are thus related to one another by their shared desire for pleasure. These are, I have argued, queer stories that hold out the potential for an as yet unrealized world saturated with pleasure and desire. Telling these stories, in itself, is an act of pleasure; imagining oneself into the place of the woman who had sex with the bear in his cave illuminates a horizon of possibility that both exceeds and expands the limits of an everyday lived world.

But the promise of these stories, I have suggested, go beyond their subversive and pleasure effects. A robust body of scholarship has traced how the philosophical boundary between nature and culture has served the purpose of naturalizing sexual and gender difference; women, as part of nature, can be brought within the ambit of culture only when purged of their animal desire. Policing women's desire, in particular, is an important part of maintaining this boundary between nature and culture, and the human and nonhuman, with all its attendant hierarchies and forms of control. What is so powerful about this genre, then, is that its promise of a new world is rooted in the sensuous mingling of human and nonhuman animal bodies. Achieving the fullness of human nature is not, in these stories, a matter of overcoming one's animal nature—a task that is held up by patriarchal ideologies as being of the utmost importance for women, whose nature is represented as perilously close to bestial—but of giving into its promised pleasures. *Bhalu ki baat* imagines another, radical, world in which these norms separating the human from the animal are undone, and the shared tug of animal desire becomes a node of relatedness between human and animal.

What is important to note, however, is that the seeds for this other world—which turns the dominant construction of sexuality, and indeed the boundary between humanity and animality, on its head—are to be found in *this* world, specifically in the relationships that emerge in the crisscrossed paths of human and bear lives. In this quotidian world, humans and bears act on one another in specific, material ways that empower symbolic engagement with the ursine world. In other words, people's everyday knowledge of the unmistakable things bears *do* as embodied beings, a knowledge gleaned through practices of living with them, is crucial to their place as the animal protagonist of these stories. How they live, how they love, what they desire—these are not inconsequential matters but are at the heart of why bears have come to be laden with such meaning in these mountains. Bringing the bear lover of these stories face-to-face with actual bears allows us to see with greater clarity how the queer futures that this genre embodies are in fact rooted in a lived present. Drawing real-life animals into our scholarly ambit illuminates how the promised pleasures of the other world invoked in these tellings can indeed be inhabited in different forms even in this world; that sensuous pleasure emerges at the intersection of bodies, both human and nonhuman. These tellings, then, are not just transgressive of patriarchal ideology, as many feminist scholars have so powerfully revealed, but also of an anthropocentric hierarchy in which the boundary between humans and animals is inked on the troubled terrain of desire. As such, what gives this genre its power is precisely the fact that it is given life in a world where human and animal bodies are porous and open to being affected by one another in ways that can produce radical, sometimes dangerous, outcomes. Perhaps *bhalu ki baat* represents another kind of otherwild, analogous to that embodied by the pig who went wild. It is indelibly marked by violence and hierarchy but also constantly undoes and troubles them.

Let me put this another way by bringing us back to a moment where bear and human lives intersected for a brief, intense moment that left an imprint on at least one of the party to that encounter. The radical possibilities of this genre are, for me, best represented by the incredible image of Gita's grandmother watching bears fuck in the forest on a day that, at first, was like any other, marked by the drudgery of having to cut firewood for the stove and fodder for the animals. She said she felt shame as she spied on them and heard their animal moans, but I suspect that she also felt curiosity and a frisson of desire as she stood concealed behind that tree for an hour, marveling at how long they could go. The image was clearly seared in her memory as she brought it to life for her young granddaughter thirty years later. Pleasure puts down deep roots. Perhaps it is in these sites of "porous pleasure" (Howe 2015) that feminist inquiry can seek the transformative potential of queer crossings like the ones attempted by these women.

Kukur aur bagh

It had been two days since Rocky, my neighbor's *kukur* (dog), was taken by a *bagh* (leopard) in the middle of the night; he was the second dog in the village to have suffered that fate since the start of the month. I knew Rocky well; he lived in the house that nestled comfortably in the curve of the main drag in Pokhri, about three hundred meters from where I lived. The first few times I met Rocky he had growled persistently while driving me away from his house. Despite the assurances of his owner, Bhaguli *bhabhi*, that Rocky wouldn't bite, I would stiffen every time I saw him charging toward me. It was only when he saw me with Lobo, the dog who belonged to the family I lived with, that Rocky warmed up to me, finally allowing me to pat his head and scratch his neck. He and Lobo would play for hours in the fields that lay between their houses, darting and feinting with such delight that even those who didn't particularly care for dogs were forced to crack a smile when watching them together. When the one female in the neighborhood came into heat, they would disappear for days on end before finally returning home together.

The day I found out Rocky had been eaten, Lobo and I passed his house in the morning while I was on my way to the next village to meet with a friend. Usually, this was where Lobo and I would part ways. But ten minutes later, much to my annoyance, Lobo had caught up with me, which meant that I would have to spend the rest of the day fending off dogs who would not tolerate his presence in their territory. In the evening, when I passed Rocky's house with Lobo still firmly attached to me, I ran into Bhaguli *di*, who warned me to get home quickly before a leopard attacked Lobo and me. She told me then that Rocky had been taken two nights ago. I was upset when I heard the news, and I asked Bhaguli *bhabhi* if she was sad (*dukhi*) about Rocky's loss. "Yes, I am very sad," she said. "He was a good dog . . . never bit anybody even

FIGURE 27. Lobo and Rocky.

though people passed in front of our house all day long. I am very sad. But your *dada* [literally, older brother; referring to her husband] is even sadder. He is very fond of cats and dogs. He was the one who brought Rocky home when he was just a puppy." "I wish there was something we could do," I said in frustration. So many dogs I knew had been eaten by leopards while I was conducting fieldwork that I had stopped keeping count.

Indeed, leopards across rural areas in India are known to thrive alongside humans because humans usually keep domestic animals, which make for easy pickings.[1] The felines in Uttarakhand were certainly no exception. Leopards were a familiar presence in the villages of Kumaon where I conducted fieldwork (Govindrajan 2015b). During the breeding season, one could hear their low moans echoing through the valley in the late evening. On most evenings, villagers hurried indoors once the sun went down and took their dogs with them; when a leopard was known to be in the vicinity, most people chained their dogs up even during the day. Despite these cautions, dogs were frequently eaten by leopards. They could be taken when they were relieving themselves in a clump of dense bushes where a leopard was lying in wait, or they could be taken from the house itself if the door was carelessly left open, as had happened to someone I knew.

Lobo himself was eaten by a leopard a year and a half after Rocky went missing. When he starting barking frantically at midnight, my friend Govind let him out thinking that perhaps he needed to relieve himself. The next thing Govind heard was a growl followed by a loud squeal before silence descended once again on the night. By the time he found a stick and ran out in pursuit of Lobo, there was no sign of him. I received the news of Lobo's death in a WhatsApp message from Govind when I was back in North America, and I spent the next few days remembering how he would dance around me on his hind legs whenever I came back home; how he would crawl into a wicker basket lined with burlap sacks to keep warm during winter nights; how he would chase the little green-backed tits who tried to pick up the soft white fur he shed every time I brushed him—perfect nest-building material; how he would run around wildly after a few play bows, stopping every now and then to look at me reproachfully out of his beautiful hazel eyes if I did not chase him. There was a lump in my throat that refused to go away when I remembered how Lobo had ushered me into his world. His death felt unreal at this distance. I dreaded returning to the mountains, where his absence would become real.

At the time of my conversation with Bhaguli *bhabhi*, however, Lobo was scratching himself vigorously while he waited for me to resume our walk home. "There is nothing we can do," Bhaguli *bhabhi* said, in response to my frustration.

> Such is the relationship between *kukur* and *bagh*. Leopards come in the form of death for dogs. Very few dogs die of disease or old age here, most are taken by leopards. This is how it has been for years. A dog's life will end in a leopard's grasp, whether the dog is young [*nainh*] or old [*budh*]. We can do nothing. Their lives have been joined in this way. If a leopard cannot find dogs [to eat] then s/he will die. A dog's death means life for them. Every time I feel sad, I think "this is their relationship and who I am to break it?"

As I hurried home with Lobo leading the way, I reflected on what it meant for two lives to be related in such a way that the life of one will end in the clutches of the other, that the death of one represents continued life for the other. Many others echoed what Bhaguli had said about the relationship between dogs and leopards—*you should not love dogs so much, or else when the leopard takes them you will be sad; treat dogs well, or else when they are taken by leopards you will feel bad; dogs are fated to die at the hands of leopards; dogs can be trusted to tell when a leopard is nearby, they know their death is close; nobody needs to teach a dog to keep an eye out for leopards, they are born knowing that this is a relationship fraught with danger.* To be born knowing that one is entangled in a relationship that eventually brings death did not seem to be a particularly good deal for local dogs, I thought to myself. But that is the tricky thing about relatedness: the terrain of intimacy it sketches, as I have argued throughout this book, is not always an easy or happy one to traverse. Whether dogs who are eaten by leopards, humans whose lives and livelihoods are at risk from wild boar protected by the writ of the state, or goats who are sacrificed by humans, to be related to another is to have one's life marked by the experience of varying levels of violence. Every act of violence only reinforces and regenerates the connection of lives and fates. Each time a dog was taken by a leopard, someone would say "see, this is their fate (*kismat*)."

On one occasion, I watched as an eleven-year-old girl weeping over the disappearance of her dog, presumably orchestrated by a leopard, was told by her mother that everyone had to leave their family someday. "Dogs must go to leopards just like you will have to leave for your *sasural* one day when you are married. Some knots (*bandhan*) are like that. Until you are with me, I will love you. But then you are someone else's." I remember thinking at the time that the analogy was a poor one. Surely a dog being eaten by a leopard had little in common with a woman leaving behind her family of birth as she journeys to her *sasural* (marital home). At the time, I thought that this was just a flippant comment, offered as appeasement for a broken heart. But in the six years that have passed since I witnessed that conversation, I have come to think that there might have been more behind the comparison between the kin relationships that are frayed and raveled by marriage and the snapping of bonds between dogs and their families when the former are snatched away from their home by leopards. In a world of multiple and sometimes conflicting entanglements, new connections must sometimes be fertilized by the break-down of old; this fact of human kinship rings true for multispecies relatedness as well. The lives of dogs, to be more specific, are related to the lives of both humans and leopards, but the full realization of one form of relatedness en-tails the violent erasure of another. For the people I knew, this complicated relatedness—enacted through violence—was only confirmed each time a dog was taken and eaten by a leopard. *This is their relationship, and who am I to break it?* This was not senseless violence, nor was it born of a simple animal instinct. This was the violence at the heart of relatedness, the expected out-come of a difficult yet inevitable entangling of lives and fates. The loss of one life, as Bhaguli *bhabhi* reminded me, is essential to the renewal of another, and connects them materially, spiritually, and affectively. As the Australian femi-nist philosopher Val Plumwood (2008, 324) reminds us in a posthumously published essay, we need to rethink our concepts of death; we are all "food and through death we nourish others." Perhaps this recursive play between life and death, regeneration and degeneration, is what ties lives together in knots of relatedness.

＊

What does it mean to live a life that is knotted with other lives for better or worse? That question, inspired by Nimmi *di*'s observation that knots of connection, once they are formed, cannot be easily disentangled has framed the scope of this book. In it, I have argued that these knots constitute a knotty relatedness, one that emerges through the myriad ways in which the potential and outcome of a life unfolds in relation to that of another. To be related to another, I have suggested, is to recognize that one's past, present, and future are gathered in theirs. This relatedness, I have put to you, is experienced by human and nonhuman animals alike, although different animals apperceive it very differently.

This is where some of you may draw the line. What I have offered, those of you might say, is an account of how relatedness is locally imagined, enacted, and negotiated by people in the Central Himalayas, an ethnographic exploration of how relatedness is extended by humans to the other-than-human in Kumaon. To be sure, in this book I have taken these interspecies encounters as constituting a kind of relatedness because that is how they were described by the people I knew. There are important political and ethical stakes to taking these claims seriously. At the time of my fieldwork and as I write this book, political and cultural debates in India around what it means to live an ethical life in relation to animals have grown only more strident. In the past few years, violent and xenophobic Hindu nationalists have attempted to narrow the contours of people's ethical imaginings around animals. To love a cow, in their worldview, is to visit violence on others in her name. To love a goat is to oppose his sacrifice even if it means snapping ties with the deities who protect both humans and animals; indeed, such deities are decried as demons and sprites unworthy of worship. To love monkeys is to render them no harm even if they commit acts of violence against you. Indeed, Hindu nationalists— and their animal-rights activist allies—have declared that only certain kinds of love for animals are legitimate. This is a political view of the world that is imposed on Hindus and non-Hindus through threats and acts of intimidation and violence.

Some of the stories I tell in this book mark their protagonists as beyond the pale of this narrowly defined ethics that is rooted in its own impulses toward violence. Asha's queer fantasies about sex with bears, fantasies that subverted the boundaries between human and nonhuman, will certainly be frowned on by these self-proclaimed guardians of a Hinduness that relies on the maintenance of women's sexual purity, believed to be the bedrock of a Hindu nation. The love that my friend Mamta felt for her cow—a love that did not prevent her from selling Chanduli to a man she suspected of being a butcher, a love that caused her to mourn Chanduli's absence—will not be acknowledged as a legitimate expression of relatedness. Bubu's tender relationship with the monkey who was sidelined by the other monkeys will be outweighed by his hatred for and violence toward the other monkeys who descended so suddenly into his world. In attempting to counter this limited understanding of ethical living, I have made two separate but related claims: first, that some acts and experiences of violence can be constitutive of quotidian relatedness, such as in the case of animal sacrifice; and second, that relatedness always entails some kind of violence, an example being the connection between dogs and leopards that I have traced in this epilogue.

I have argued in this book that violence does not always preclude care, attention, and even love. I was witness, of course, to acts of horrifying and nonchalant violence, such as the occasion when Devrani, an abandoned female dog that I had helped raise, was poisoned, by whom I still do not know. I do not believe that the violence of the death imposed on her was mediated by love, care, or concern. But there were other moments, as I have explored throughout this book, when I witnessed people trying to enact ordinary ethics from the smoldering ashes of violence. I am reminded here of the observation made by the anthropologist Robert Desjarlais in his work on experiences and philosophies of death among the Hyolmo. "Many deaths are violent," Desjarlais (2016, 48) writes; "There is nothing gentle or graceful about them. Hyolmo people know this to be the case. . . . 'Life is like this,' they might say. And yet this understanding does not prevent them from trying to provide gentle, calming care to others, when they can, and it does not deter them from preparing for a good death." Even in the more complicated moments that I have traced—in the cases where responsibility for a violent death was easily assigned—many people in Kumaon tried to atone for their actions in multiple ways: by providing gentle and loving care; by remembering that their continued presence in the world owed itself to the deaths of others; by acknowledging guilt and remorse for their actions.

Perhaps you might read my attempt to reconcile love and violence as rooted in a naive and pernicious humanism that privileges the remorse of the human over the life of the animal. I have no easy response to that accusation. I am painfully aware that we, as humans, must account for what Alice Kuzniar calls our "disavowal of identification" with animal life (Kuzniar 2006; cf. Garcia, forthcoming). However, I remain simultaneously cognizant that the disavowal of animals intersects in complicated ways with the disavowal of some humans in contemporary India and that there is no solace to be found in pure and uncomplicated categories such as the "'human" or the "nonhuman."

To put it another way, the promise of posthumanism must engage the lessons of postcolonialism and vice versa. It is this recognition, then, that leads me to believe that the path toward kin making that I have illuminated in this book—a path that is challenging and almost disappears in some parts but the traversal of which is also rewarding—contains within it the possibility for a world of mutual recognition, avowal, and response between different beings.

But what do we make of love given that so many—from *gau-rakshaks* to animal-rights activists to ordinary villagers—claim to be motivated by it? Can love itself be an absolution?[2] No. All love is decidedly uninnocent, whether it is the *gau-rakshak* Pratap Pant's love of the cow or the love that my friend Mamta felt for her cow Chanduli. Both constitute forms of affective attachment that ultimately privilege the subjectivity and desires of the self over that of the other. In other words, love does not entirely undo the subject's interest in herself. Lauren Berlant's (2011b) observation on the relationship between love and ethics is worth invoking here. Berlant comments that love is not entirely ethical because it is always related to desire. "Many kinds of interest," she writes, "are magnetized to the rhythm of convergence we call love" (Berlant 2011b, 685). Naisargi Dave (2015) takes this argument a step further when she writes that "love is an injustice because when we love it is the one or ones who are special to us that we save." Both the *gau-rakshak* and people like Mamta love with interest and through injustice. For *gau-rakshaks* like Pratap Pant, the love of cows is bound up in their desire to create a sovereign, masculine Hindu nation. Mamta's love of Chanduli was tethered to and emerged out of practices of care that were directed at ensuring the latter's productivity for human ends.

And yet there are critical differences between these two kinds of love, both in the interests that motivate them and in what they do to the person who feels them. In other words, these different kinds of love were constituted by and constituted different kinds of politics. Mamta's love emerged in the crucible of a rural political economy in which women are almost entirely responsible for the tedious and arduous labor of caring for animals. This gendered labor was crucial in creating the terrain on which affective multispecies attachments materialized and flourished. Mamta's was a *situated* love that grew out of her everyday embodied entanglements with Chanduli. The love that *gau-rakshaks* like Pratap Pant bore for cows, on the other hand, was more transcendental in nature, invoking the status of the cow as the mother of the Hindu nation as the reason for its existence. Unlike Mamta, his attachments were not to singular individuals but to the gendered metaphor of the cow that could serve as the locus of a violent religious nationalism. These differences are important because even though love itself is a troublesome category, it matters immensely where one is situated within the terrain on which it emerges.

Its origins apart, love also did very different things to people like Mamta and Pratap Pant. For Pratap Pant, love was an unambiguous force. His love for the cow confirmed his politics and had no room for uncertainty, remorse, or guilt for the particular ways in which that love played out. Mamta's love, on the other hand, left her open to experiencing grief, ambivalence, and even shame. This questioning of the self, no matter how brief, interrupted any certainty about what it meant to live well and flourish together. It turned the matter of ethics from a transcendental and morally unequivocal philosophy into an ongoing, ambiguous, and always incomplete *doing*. Bhrigupati Singh and Naisargi Dave remind us that ethics are a question not just of "how to live" but also of "how to kill" (Singh and Dave 2015). One of the questions an anthropology of ethics must engage, they write, is whether the inevitability of nonhuman killing means that it no longer matters. It is worth noting that the inevitability of Chanduli's death did not make it matter any less for Mamta. Perhaps it only made her death matter more because of the ways in which it forced Mamta to articulate why it was inevitable and to confront her own responsibility in declaring its inevitability.

Again, Lauren Berlant's series of provocations on love help us to make sense of the ethical mess that love made for Mamta. Berlant reminds us that the power of love should be sought in its interruption of sovereignty.[3] Love, she says, causes the one who feels it to be caught up in a "rhythm of an ambition and an intention to stay in sync" with another (Berlant 2011b, 683). Love's demand of synchronization means that one gives up a little part of oneself to be with the beloved, thus rupturing the sovereign boundaries between bodies. It was love that compelled Mamta to dwell in the ambiguity and contradictions of violence. Even if love did not entirely undo Mamta, it was a risk that opened her up to new kinds of relatedness and thereby to new kinds of ethics.

In reflecting on the ethical power of the kind of love that moves one, I am not advocating that we surrender ethics to the "hegemony of love" (Povinelli 2006, 25). I agree with those who note that love is not entirely ethical. Yet I am not convinced that love is entirely lost to us as the basis of a different world. Love is not equivalent to ethics, to be sure, but it does not dwell entirely outside the realm of the ethical either. Love might not be perfect, and it is certainly not enough to singlehandedly transform the worlds we inhabit. But love can sometimes hold open the fleeting possibility of another world. While it cannot answer ethical questions, it can pose them anew in ways that allow people to temporarily step out of themselves and be open to being transformed by others. Love is often unjust, yes, but it can also be "properly transformational" in that it provides us "with the courage to take the leap into a project of better relationality" (Berlant 2011b, 690). Love, as Berlant puts it, is change "without guarantees, without knowing what the other side of it is, because it's entering into relationality" (Berlant 2011b, 690). Love, with all its troubles, then, ties us up into knots of relatedness that can often be uncomfortably tight but that also invite us to dwell a little longer in the possibility of how we might be, together, differently.

Is love restricted to the human? Throughout the book I have argued that love and relatedness, which encompasses love but also its affines and antitheses, constitute the terrain on which animal being and becoming unfold. Is it anthropocentric to argue that nonhuman animals try to live "in sync" with other beings, that they have a sense that their life is tightly interconnected with other lives? Perhaps, but as Brian Massumi says it is a risk worth taking "in the interests of following the trail of the qualitative and the subjective in animal life, and of creativity in nature . . . with the goal of envisioning a different politics . . . freed from the traditional paradigms of the nasty state of nature and the accompanying presuppositions about instinct permeating so many facets of modern thought" (Massumi 2014, 2). Perhaps, Massumi continues, in the course of such investigation, "we might move beyond our anthropomorphism as *regards ourselves*: our inveterate vanity regarding our assumed species identity, based on the specious grounds of our sole proprietorship of language, thought and creativity."

This book has been inspired by exactly this hope—that a productive and open anthropomorphism that begins with the awareness that not all chasms are bridgeable can challenge our hubris about the distinctiveness of human animals. As I learned more about the lives and preferences of the specific animals—both human and nonhuman—that I lived with in Kumaon, I saw that they shared a broad sense that the potentialities of their lives were connected to that of another even though that connection might have been recognized or experienced in very different ways. While the kid who followed Bimla *chachi* around might not exactly have thought of her as a "second mother"— the term *chachi* used—she certainly seemed to recognize in her what Donna Haraway terms "the presence of a significant other," someone whom she could play with and love, someone that she could trust to take care of her. When leopards move into human-dominated areas, they do so with the expectation, perhaps even the knowledge drawn from past experience and from others of their kind, that they will thrive in these zones of multispecies copresence. They recognize that they flourish in proximity to humans, not at a distance from them, even though proximity can often be a double-edged sword. When Chanduli, Mamta's cow, began to bellow when the Tempo that she was in drove away, I believe she had a premonition that her life was about to change; her frantic cries were an unmistakable expression of anxiety and possibly even grief. She might not have known that this separation would be permanent, but she certainly did seem to recognize that the intimacy of her everyday relationship with Mamta was about to be disrupted. This was animal instinct, which, Brian Massumi notes, relates all animals to one another. A radical animal politics, he so powerfully reminds us, "has no fear of instinct."

The stakes of this openness to the mutuality between humans and animals are more important than ever. My commitment to the idea of relatedness, with all its promises and perils, is motivated by ethical and political concerns. As the idea that the world is in the clutches of the Anthropocene—the era in which humans act as a geological force—continues to storm through the academy and public life with astonishing force, it is difficult to deny that humans have joined the ranks of other beings in producing effects that are planetary in scale and kind. Several powerful critiques of the Anthropocene are worth recalling here—the homogeneity of the category human, which overlooks how both responsibility for and vulnerability to ecological collapse are unevenly distributed; the hubris enshrined in the assumption that humans work alone without assistance from a whole host of other species and inanimate actors, even when the work they are doing is destructive; the reminder that humans modified their environment long before the period when the Anthropocene was instituted.[4] To be sure the Anthropocene is a "gift armed with teeth" (Howe and Pandian 2016).

However, it is also increasingly clear in this moment when the star of "mass man-made death" hangs over us that we must remake old alliances and make new ones if we are to sustain regenerative life (Bird Rose 2011). Donna Haraway (2016, 100) draws on the work of Anna Tsing to suggests that the Anthropocene is a period in which "places and times of refuge for people and other critters" have been destroyed, leaving us in a world "full of refugees, human and not, without refuge." If we are to create spaces and times of "ongoingness" and "flourishing," Haraway says, we must make kin. I concur on the urgency of making kin, on the need to "make-with, become-with, compose-with" and, indeed, to decompose-with (Haraway 2016, 102). It is my hope that this book, by showing how people and animals in a particular place and time make and experience relatedness with one another, has illuminated small paths toward (re)creating refuges for the human and the nonhuman or other-than-human to become in relation to one another. With all its possibilities and perils, relatedness might be all we have.

Notes

Chapter One

1. *Di* is a diminutive for *didi*, older sister. Throughout the book, I use this and other kinship terms, especially *da* (brother), *chacha* (father's younger brother), *chachi* (father's younger brother's wife), *bhabhi* (sister-in-law), *mama* (mother's brother), and *mami* (mother's brother's wife) to describe some of the people who figure in the book. The names themselves are pseudonyms except in the case of well-known public figures.

2. *Moh-maya* is often translated as an illusion. However, as in other parts of India, *maya* is used here to signal the affect that binds humans (and nonhumans) to one another. As Sarah Lamb points out, "maya not only consists of what we would classify as emotional ties but involves substantial or bodily connections as well. People see themselves as substantially *part* of and *tied* to the people, belongings, land, and houses that make up their personhoods and lived-in worlds" (Lamb 2000: 116).

3. *Radhika* is a diminutive of Radha, Krishna's beloved.

4. David Schneider's landmark book *A Critique of the Study of Kinship*, published in 1984, marked an important shift in the terrain of kinship studies from the biological to the cultural. Schneider (1984, 165–66) criticized anthropological models of kinship that claimed to have discovered kinship "out there" as it "actually is." Anthropologists, he argued, needed to distance themselves from conceptual models of kinship as biological being and focus instead on the daily "doing and performing" of kinship within particular ethnographic locations. Schneider's emphasis on kinship as a doing, to be located in the practices of everyday life, was taken up by a number of scholars, including the anthropologist Janet Carsten (2000, 3) who offered the term *relatedness* to describe "indigenous statements and practices" of what it means to be related, "some of which may fall quite outside what anthropologists have conventionally understood as kinship." In particular, Carsten argues, exploring what it means to be related in particular places allows us to understand the complicated ways in which the biological and the social interact in everyday life in ways that do not conform to the pregiven analytical opposition between biology and culture in studies of kinship that were rooted in Euro-American epistemologies. Such critiques of early studies of kinship as rooted in European epistemological and ontological categories were, as Franklin and McKinnon (2001, 3) note, "exemplary of a broad shift within anthropology in the 1970s and 1980s toward more self-critical and reflexive approaches . . . and a rejection of objectivist models in favor of more hermeneutical ones."

5. For a wonderful account of the role of Hanuman in South Asian culture and religion, see Lutgendorf 2007.

6. A number of anthropologists increasingly argue in favor of understanding human relationships with the other-than-human —whether animals (Bird Rose 2011; Cassidy 2002; Cormier 2003; Jalais 2010; Weaver 2015), gods and spirits (Ramberg 2014; Singh 2015; Taneja 2017) or plants (Archambault 2016; Besky 2013)—in terms of kinship.

7. The phrase "domestic but not domesticated" is inspired by Stacy Alaimo whose book *Exposed* (2016) "considers the pleasures of inhabiting places where the domestic does not domesticate and the walls do not divide" (1).

8. In the villages where I worked, very few people kept buffalos or chickens. Buffalos required more water and fodder than was easily available, and chickens were considered dirty and diseased by many, including the handful of families who experimented with keeping them on a large scale.

9. http://www.ivri.nic.in/campuses/mukt/default.aspx.

10. For more on how people in South Asia are formed as persons through transactions with one another and the environment, see Marriott 1976; Daniel 1984; Raheja 1988; and Ramberg 2014.

11. As Jonathan Parry notes, the idea put forth by Marriott and others that Indian thought is highly monistic and portrays the self as divisible and constantly changing is "somewhat overdrawn." How does the rigid order of caste, Parry (1994, 114–15) muses, square with the idea that persons are made up of a "transformable bio-moral substance which is continually modified by transactions in which they engage"? He concludes that Indian thought is characterized by a complex mingling of both dividualism and individualism. See also Lamb 2000.

12. One of the troubling legacies of the Chipko movement is the fetishization of mountain villagers in Uttarakhand as protoenvironmentalists. The Chipko movement began in 1973 when a group of villagers in Garhwal clashed with the Uttar Pradesh Forest Department over its decision to permit a sporting goods company to chop down hundreds of trees. When workers arrived to chop the trees, the villagers—primarily women—clung to the trees (*chipko* in Hindi means to stick) and prevented felling. These protests spread throughout the Garhwal Himalayas and soon received widespread academic, activist, and media interest (Guha 1990). Environmentalists and academics argued that villagers, in preventing the felling, had recognized that development would destroy their environment and sociality (see, e.g., Escobar and Alvarez 1992); some argued that the women protestors, in particular, had revealed that they shared a close connection to nature (see, e.g., Shiva 1988). However, more contemporary scholarship on Chipko argues that such narratives elide the complex motivations of protestors who were not antidevelopment per se but wanted greater economic control over forests (Agarwal 1992, 1994; Linkenbach 2007; Mawdsley 1998; Rangan 1997, 2000). Despite this complicated history, mountain villagers are often hailed by environmental activists as indigenous stewards of nature who display a powerful ecological consciousness. For a powerful account of the deleterious effects of this representation of indigenous people as "eco-savages" who live in harmony with nature and worship it, see Shah 2010.

13. Anthropology's fetishization of villages as timeless and pristine sites, a legacy of its own colonial entanglements, has been rightly and roundly criticized over the last few decades. For an overview of the debates over "village studies" in South Asia, see Mines and Yazgi 2010.

14. See, e.g., the work of Eduardo Viveiros de Castro (2004, 2012, 2015). For a discussion of the continuities between the use of culture and the use of ontology in anthropology, see Carrithers et al. 2010 and Graeber 2015. On the unacknowledged relationship between the ontological turn and the work of indigenous scholars, see Todd 2016. For a review of the anthropology of ontology, see Kohn 2015.

15. In the wake of the devastating floods and cloudbursts that accompanied monsoon rains in 2013, *paharis* were quick to point out that the aggressive damming of rivers, large-scale deforestation by the timber mafia, and the illegal mining of riverbeds for stone had played an important role in exacerbating the loss of life and property that was caused by the rains. All of this, they argued, had taken place with the active support and collusion of the state, which had neglected to sustainably and equitably develop the region.

16. Quoted in Atkinson 1973.

17. The anthropologist Naisargi Dave, who worked closely with Gandhi in the course of her research on animal-rights activism in India, notes that Gandhi works "seven days a week from morning to night . . . advocating for cows, pigs, dogs, cats, donkeys, and chickens by writing weekly newspaper columns, lobbying fellow parliament members, and threatening over the phone to have people beaten, hanged upside down in their underwear, killed, maimed, and disappeared. Given her history, nobody takes these as idle threats" (2014, 438–39).

18. For more on the history of reform and revivalism in India, see Fuller 1992. More specifically, on the reformist campaign against sacrifice in Uttarakhand, see Govindrajan 2014.

19. On human entanglements with nonhumans and multispecies ethnography more specifically, see Baynes-Rock, 2015; Blanchette 2015; Dave 2014; Desmond 2016; Haraway 2008, 2016; Kirksey and Helmreich 2010; Kohn 2013; Kosek 2010; Locke 2017; Nading 2014; Parreñas 2012, 2016; Porter 2013; Raffles 2002; Van Dooren, Kirksey, and Münster 2016; and Tsing 2015. These engagements in the discipline of anthropology are inspired by and in conversation with different perspectives on the relationship between humans and nonhumans, particularly new materialism and actor-network theory. On the agency, vitality, and animacy of matter, see Bennett 2010; Coole and Frost 2010; Chen 2012. On actor-network theory, which sees social action not in terms of the intentionality of actors but as the outcome of networks of association between a host of different actants, see Latour 2005. For an overview of multispecies perspectives in critical animal geography, see Gillespie and Collard 2015. On the multispecies turn in history, see Ritvo 2004, 2007; and Mikhail 2011, 194–96.

20. https://culanth.org/articles/139-twenty-five-years-is-a-long-time.

21. There are a number of different semantic categories that one can use to indicate the caste status of individuals and groups. The most precise and narrow category is *jati*, a polysemic term that can be roughly translated as "caste," "kind," or "species." Different *jati* groups within and across *varnas* are distinguished from one another by restrictions on intermarriage and the sharing of food. Kumaonis usually classified *jatis* into *thul* (high) and *neech* (low). As Bhrigupati Singh (2015, 27) notes in his work on Rajasthan, "rather than being disavowed for their evolutionary overtones," such terms are "accepted, desired, and even demanded by various social groups in India in recent decades." This is because legal classification as a Scheduled Caste gives groups who face ongoing oppression within the caste system access to education, jobs, and a range of affirmative action programs that are guaranteed under the Indian Constitution as well as some measure of protection from upper-caste violence through laws such as the Scheduled Castes and Scheduled Tribes (Prevention of Atrocities) Act, 1989. The lower-caste groups in the villages I worked in were all eligible for Scheduled Caste (SC) status as members of Shilpakar, that is artisan or craftsman, *jatis*. However, very few people belonging to these *jatis* referred to themselves using this legal nomenclature in everyday conversation unless they were describing specific contexts in which they needed to establish their SC status; e.g., someone might say that they needed their SC certificate to seek admission to a college or apply for a government job. On the whole, lower-caste individuals identified themselves as Shilpakar, Arya *log* (people), or Ram *log*. A few lower-caste villagers also used the term *Harijan* (literally, people of god) to identify themselves. The term

Harijan was coined by Gandhi in the 1930s in an effort to find a nonstigmatizing word to refer to lower castes, particularly "untouchables"; incidentally, he translated the word as "children of god." However, the term is considered pejorative by many for the ways in which it smacks of pity and condescension. In Uttarakhand, a number of upper castes also used the term *Harijan* as well as Ram/Arya *log* to describe lower-caste people. In some contexts, they also often used the word *Dom/Dum* for "*neech*" *jatis*, which is widely held as a term of caste abuse by lower-caste groups. The term was used mostly behind the backs of those to whom it referred, except during heated fights when it could be wielded as a demonstration of caste power. The term *Dalit* (oppressed), which is widely used elsewhere in India and is a term I prefer to use when referring to lower-caste groups, had little valence in the region; almost no lower-caste individuals I knew used it to identify themselves despite knowing well its political and social history. Given this complicated semantic history, I use a number of different caste terms depending on the context. Given their power-laden histories, I use *Harijan* and *Dom/Dum* only when quoting people who used the term while talking to me. Where possible, I use the specific *jati* name when referring to an individual or group. People's names also index their *jatis*; for instance, Ram, Tamta, and Arya are always lower-caste surnames; Bisht, Rawat, and Nayal are *Thakur* surnames; and Bhatt, Guthuliya, and Pandey are *Pundit* surnames. Elsewhere, I use the terms *upper caste(s)* and *lower caste(s)*, *Pundit*, *Thakur*, *Dalit*, formerly "untouchable," or Scheduled Caste when referring to groups or individuals who bear that legal status. For a historical account of caste relations in Kumaon, see Joshi 2015.

Chapter Two

1. In a previous article (Govindrajan 2015a), I mistakenly dated this visit to 2010, the date of a previous trip that I had made to the same temple.

2. Ma Kalika is the *isht devi* of the Kumaon regiment of the Indian Army. There are several stories that are told to explain why she has this exalted position. In one tale, she supposedly came to the commander of the regiment in a dream to warn him of an invasion by the Chinese army. In another, she is believed to have saved soldiers of the regiment who were on a sinking ship. Large sections of the temple complex have been built or renovated by the army.

3. On possession in ritual contexts in Uttarakhand, see Malik 2016 and Sax 2009.

4. The term *gotra* refers to a group that includes all persons who trace their descent in an unbroken male line from a common male ancestor.

5. Date of Judgment: 19.12.2011. "Coram [*sic*]": Hon'ble Barin Ghosh, C. J. (chief justice) and Hon'ble U. C. Dhyani, J (justice), emphasis mine.

6. *Bali* is the commonly used term for animal sacrifice in Hindu discourse. As Jonathan Parry (2008, 242) points out, "*bali* refers both to sacrifices voluntarily offered and to the victims seized by disappointed deities who have been denied their due." The term *puja*, in Uttarakhand, is also used to refer to animal sacrifice that is voluntarily offered but denotes any kind of offering to deities more broadly.

7. Such rituals often involve sacrifices that are piacular or expiatory (Evans-Pritchard 1954) in nature. Sacrifices that are intended to heal those who are afflicted by the "bad sacred" (Parry 2008) are offered not only to angry deities but also to ghosts, or the spirits of unhappy ancestors. These sacrifices can be presented at temples but are more often offered during the middle of the night around streams where demons such as Masan (lord of the cremation ground who appears in many different forms) are said to live or at the place where the victim is thought to have been seized by the afflicting ghost or spirit. As a woman, I was not allowed to attend these nightly

rituals on the grounds that I, too, would be seized by Masan or a spirit. As such, this chapter focuses largely on *khushi pujas*. For an account of expiatory *pujas* in Uttarakhand, see Sax 2009.

8. As Evans-Pritchard (1956) points out, sacrifices to minor deities are usually offered as ransom, while sacrifice offered to major deities is in the nature of a gift.

9. However, as Veena Das (1983, 457) points out, substitution is a complicated matter. For instance, in Vedic sacrifice "it is not correct to say that the prescribed material is always to be preferred to the substitute. For example, a man wishing for riches may perform the *hiranya* sacrifice. In this a hundred gold pieces . . . have to be boiled and offered and the mode of offering follows the procedure for boiled rice. So it is apparently that the boiling of the gold pieces is a substitute for the act of boiling rice, and yet since the sacrifice is geared to a desire for an object, it is the substitute which is to be preferred."

10. In this work, Evans-Pritchard (1954) is careful to situate the relationship between sacrificer and sacrificial victim in the broader context of social life. He observes that sacrificial equivalence, the idea that two beings are of the same order and can therefore be substituted for one another in sacrifice, had always been between a man and an ox. This equivalence, Evans-Pritchard insists, can be understood only in terms of the broader identification of the Nuer with their cattle in daily life. The use of ox names for men, the place of oxen in rituals of initiation, the value attached to cattle for their milk and as bridewealth are all emblematic of the identification of men with cattle in sacrifices. However, despite his focus on the close relationships that the Nuer shared with their cattle, ultimately, for Evans-Pritchard, it was not the particular "ox of initiation" but rather the "idea of oxen" with which people identified. He writes, "Any ox will therefore serve the purpose, or, indeed, no ox at all—only the memory, or rather the idea, of an ox" (Evans-Pritchard 1954, 185).

11. Lévi-Strauss was particularly confounded by the substitution of a cucumber for an ox in the Nuer sacrifices described by Evans-Pritchard. He writes that sacrifice "permits a continuous passage between its terms: a cucumber is worth an egg as a sacrificial victim, an egg a fish, a fish a hen, a hen a goat, a goat an ox. And this gradation is oriented: a cucumber is sacrificed if there is no ox, but the sacrifice of an ox for want of a cucumber would be an absurdity" (Lévi-Strauss 1966, 224).

12. For an overview of these critiques, see Keenan 2005.

13. Das (1983, 456) points out that in Mimamsa theories of sacrifice, the five categories of domestic animals who can be sacrificed include man, horse, ox, sheep, and goat. There are strict rules governing the substitution of one sacrificial victim for another, and, in some kinds of sacrifice, no substitution is possible. One element of the natural world, in other words, cannot seamlessly stand in for the other.

14. The holiness of the animal, Smith argued, was reflected in the fact that his life could not be taken by any one individual and that he could be sacrificed only through the consent and participation of all the members of the clan.

15. Nadasdy (2007, 26) argues that scholars of human-animal relations such as Tim Ingold "shy away from some the most radical implications of [their] own work and . . . [treat] certain important aspects of northern hunters' conceptions about human-animal relations as just metaphor, including the notion that animals are thinking beings who might consciously give themselves to hunters."

16. On the intersection between gender, political economy, and development discourse in Uttarakhand and the Himalayas more generally, see Agarwal 1994; Berry and Gururani 2014; Dyson 2014; and Gururani 2002.

17. And yet, some anthropologists point out that despite the grueling nature of this routine,

women can also experience this labor as creating an opportunity for unsupervised and joyous female sociality and a respite from the household (Dyson 2014).

18. Globally, care work is not only gendered but also racialized and classed. Care economies are increasingly constituted by the intersection of gender, race, and class inequality. See, e.g., Parreñas 2001.

19. On affect, see Berlant 2011a, 2011b; Dave 2012; Massumi 2002; Mazzarella 2009; Parreñas 2012; and Rutherford 2016.

20. On embodied language, see Despret 2013; Govindrajan 2015a; Haraway 2008; Parreñas 2012; and Weaver 2013.

21. See also Ben Campbell's (2012) ethnographic account of the affective relationships that Tamang people in Nepal share with their livestock. Campbell (2012, 203) writes that "living in closely dependent domestic proximity with animals generates all kinds of emotions. . . . It is not domination pure and simple, but a convivial, domestic distributional hierarchy based on practices of interpersonal attentiveness, that characterizes this milieu."

22. https://thomvandooren.org/2013/11/02/keeping-faith-with-death-mourning-and-de -extinction/.

23. In his work on rituals of affliction and healing in Uttarakhand, William Sax (2009, 162) describes a *hamkar puja* (an expiatory ritual) in which the *guru*, Jagdish, does a separate *puja* after the sacrificial goat has been killed. He writes, "after the sacrifice, Jagdish does the *puja* of the head, in which a candle is placed on the head of the sacrificial victim, a prayer is said for its *atma*, and the candle is blown out. All gurus do this ritual: some say it is for the sake of the animals, others that it is to eliminate the guru's sin at being involved with animal sacrifice. Jagdish and his brother Mohan always do the ritual very quickly; not, says Jagdish, because of any danger, but rather so that the animal can quickly obtain *moksha* or liberation".

24. See Faubion 2011 and Laidlaw 2014 for critiques of Zigon's sharp distinction between the everyday and the moments of social crisis.

25. On contemporary debates around sacrifice in Jharkhand and Rajasthan, see Shah 2010 and Singh 2015, respectively.

26. In his powerful and provocative work on anti-Muslim pogroms in the Hindu-dominated Indian state of Gujarat, Parvis Ghassem-Fachandi (2012, 12) makes the point that *ahimsa* began as a "protective technique against the effects of the necessary violence" in the ritual of sacrifice. However, contemporary forms of Hindu nationalism in the state, while advocating ahimsa toward the animal victims of sacrifice, have replaced the original animal victim of religious sacrifice with a Muslim victim. The disassociation of sacrifice and meat eating from Hinduism has, as Ghassem-Fachandi (2010, 161) notes elsewhere, disturbing implications. In Gujarat, Hindu agents of violence were able to argue that they were simply doing to Muslims what Muslims were identified as doing—slaughtering animals and consuming meat.

27. Gauri Maulekhi, interview by author, September 2010.

28. "Should Bakri-Id Also Be Celebrated in Eco-Friendly Manner Like Other Festivals?," *Zee News*, September 12, 2016, https://www.youtube.com/watch?v=m-MW1L8sQwE.

29. For a masterful account of buffalo sacrifice and its symbolism in Tantric Hinduism, see Biardeau et al. 2004. On buffalo meat and ritual pollution, see Berreman 1963; Sax 2009; and Hiltebeitel 1980.

30. In 2010, the PFA filed a police report against the Devidhura Temple Committee for permitting buffalo sacrifice. The administrators of other temples in the region were afraid that similar cases would be filed against them as well.

31. In Dewan Singh Bhandari & Others vs State of Uttarakhand & Others (Writ Petition No.

1898 in the High Court of Uttarakhand, 2012), the petitioner requested the court to allow sacrifice at a temple in Bageshwar district where, they claimed, it had been a "thousand-year tradition." To satisfy the court that the animal would be killed to provision food, the petitioner assured the court that the animals would be consumed by people after they were killed.

32. Aditya Malik notes that Golu's divine sovereignty often rivals that of the postcolonial state. Several people told him that Golu's *darbar* (his court) was more powerful than a civil court and that his justice was unquestionable. Malik (2016, 5) argues that the extent of Golu's juristic power in Kumaon reveals how "multiple systems of justice exist simultaneously in an overlapping, merging, and yet sometimes oppositional manner in the same social, political, and cultural space." The High Court's decision on sacrifice sparked a moment of opposition between these two sovereign regimes, causing many to question and subvert the authority of the postcolonial state.

Chapter Three

1. The gendered nature of this kinship metaphor is striking. Gupta (2001, 4296) argues that in the nineteenth century the cow was linked with "building a strong nation, a nation of Hindu men who had grown weak and poor from lack of milk and ghee. For a body of healthy sons, cows became essential. . . . The material body of the mother cow was equated with [*sic*] Hindu nation, where she was the benevolent mother, whose womb could provide a 'home' to all. Like a mother, she could feed her sons with milk, making them stronger."

2. On Hindu nationalism and its history of cow protection, see Engineer 1995; Hansen 1999; Jaffrelot 1996, 2007; and Van der Veer 1994.

3. One could argue, on the other hand, that this failure to eliminate cattle slaughter is not necessarily a failure in the ideological sense in that it allowed right-wing organizations to claim a continued state of emergency in which Hindus and their mother, the cow, had to be protected from threats to the nation at all costs.

4. The smuggling of cows from India to Bangladesh has become an explosive political issue in recent times. In 2015, during a trip to the Indian state of Meghalaya, which shares a border with Bangladesh, I ran into a group of men on the highway who were in charge of about sixty ageing bulls. When I asked them where they were taking the bulls, they responded that the bulls were going to Bangladesh. When I inquired as to the provenance of the bulls, they said the animals had come from Rajasthan, Punjab, Uttarakhand, and West Bengal. "They come from all over India," one of the men said. The PFA, *gaurakshaks*, and ordinary citizens disturbed by the porosity of the border are demanding strict police and army action against the smuggling of cattle. In 2015, the supreme court of India responded to a PIl filed by the Akhil Bharat Krishi Go Sewa Sangh and asked the central government to take steps against the smuggling of cattle across India's borders. It also asked individual states to check the movement of cattle outside state borders. "Govt Takes Steps to Stop Cow Smuggling to Bangladesh," *Livemint*, September 5, 2016, http://www.livemint.com/Politics/pTfDFlDozvl1U1zGORqhIN/Govt-takes-steps-to-stop-cow-smuggling-to-Bangladesh.html.

5. "Animal Welfare Activists Blame State Govt for Cow Slaughter Cases," *Daily Pioneer*, September 3, 2013, http://www.dailypioneer.com/STATE-EDITIONS/dehradun/animal-welfare-activists-blame-state-govt-for-cow-slaughter-cases.html

6. "PFA Raids Cattle Market; Case Registered," *Tribune*, June 12, 2012, http://www.tribune india.com/2012/20120613/dplus.htm#8.

7. "VHP Finds Cash Cow: Cosmetics," *Ahmedabad Mirror*, October 27 2010, http://epaper

.timesofindia.com/Repository/getFiles.asp?Style=OliveXLib:LowLevelEntityToPrint_MIRROR NEW&Type=text/html&Locale=english-skin-custom&Path=AMIR/2010/10/27&ID=Ar00100.

8. This is in keeping with the caste dynamics of the state, the population of which, as discussed in the introduction, is largely upper caste.

9. See, e.g., Chigateri 2011.

10. On incommensurability, see Povinelli 2001.

11. *Shakti* is a term that refers to the "immanent, manifest and acknowledged energy or power of the divine," usually a feminized divine (Pocock 1973, 88).

12. Christophe Jaffrelot (2008, 3) argues that a Hindu nationalist culture of outrage took shape in the late nineteenth century as a reaction to a strong sense of vulnerability. Hindus, though the numerical majority, saw themselves as weak compared with Muslims because of their inner divisions along caste and sectarian lines. This "majoritarian complex of inferiority made Hindu nationalist leaders prompt to outcry as soon as some of their sacred identity symbols were 'under attack' because of religious minorities, be they Muslim or Christian." Cow-protection movements, Jaffrelot argues, exemplify this pattern.

13. De 2013.

14. "Cops Told to Prevent Illegal Slaughter during Bakri-Id," *Daily Pioneer*, October 20, 2012.

15. Gauri Maulekhi, "Ban on Cattle Sale for Slaughter: Can We Stop Outraging and Focus on Regulating Cattle Markets?," *Firstpost*, June 1, 2017, http://www.firstpost.com/india/ban-on -cattle-sale-for-slaughter-can-we-stop-outraging-and-focus-on-regulating-animal-markets -3503673.html.

16. http://www.gaylaxymag.com/latest-news/anti-gay-modi-stickers-ask-people-to-choose -between-cow-protection-and-gay-protection/#gs.QZdkotw.

17. For more on the relationship between Hindu nationalism, queerness, and masculinity, see Bachetta 1999; Dave 2012; and Hansen 1999.

18. On the social profile and motivations of *gaurakshaks*, see the excellent profile by Aman Sethi and Adi Prakash, "How to Become a Gaurakshak Celebrity," *The Quint*, Politics, August 11, 2016, https://www.thequint.com/politics/2016/08/11/how-to-become-a-gaurakshak-celebrity -narendra-modi-una-dalit-muslim-uttar-pradesh-muzaffarnagar.

19. In protest, Dalits across the state declared that they would no longer pick up or skin dead cows. In a symbolically powerful gesture, they threw cattle carcasses into various public offices and suggested that *gaurakshaks* display their self-proclaimed love for the mother cow by disposing of the corpses. Meera Velayudhan, "Dalits and the Holy Cow," LeftWord blog, (August 26, 2016), http://mayday.leftword.com/blog/dalits-the-holy-cow/.

20. A number of other Muslim rulers took care to respect the sensibilities of their Hindu subjects when it came to cows. Hindu discomfort with cow slaughter prompted a number of Mughal rulers such as Akbar and Bahadur Shah Zafar to ban the practice in parts of their kingdom. Christopher Bayly (1988, 187) notes that when the Mughals liberated Delhi in 1857, they "banned the Muslim sacrificial practice of cow-slaughter . . . on the grounds that it might offend the 'eastern [i.e. Hindu Brahmin and Rajput] soldiers.'"

21. This judgment was based on Section 295 of the Indian Penal Code, which declared it an "offense to destroy an object held sacred by any class of persons."

22. See also McLane 1977.

23. There is a rich literature on cow-protection riots in the nineteenth and early twentieth centuries. See, e.g., Pandey 1983; Yang 1980; and Gould 2012.

24. Here I am not opposing belief as an internal state of reflection to politics as external

action or arguing that belief is untainted by politics. However, while recognizing how ideas of religious belief and truth are formed in the crucible of history, culture, and power, I am wary of the claim that power *creates* religious meaning. Belief, even as it remains a highly contested term (see, e.g., Ruel [1982] 2002; Ewing 1994), cannot be reduced simply to an artefact of power, a subset of politics.

25. As Cassie Adcock (2010, 300) notes, "arguments concerning the efficacy of the cow as a 'religious symbol' tend to reproduce false assumptions that have a deep colonial legacy, such as 'the religious character of the mass of the people, or the monolithic unity of faith' in India." See also Chatterji 1994.

26. E.g., Bhrigupati Singh (2011, 444) argues that the theological significance of the cow in the part of Rajasthan where he worked is "a marker of the peculiar response . . . to the global moral problem of the ascetic ideal, its absorption into the everyday life of householders (through vegetarianism, teetotaler practices etc.), and the turning away from human-animal sacrifice."

27. Amul, an acronym for Anand Milk Union Limited, is a dairy cooperative that is the largest producer of milk in India. This cooperative was set up in the district of Anand in the state of Gujarat in 1946, a year before India's independence, and is mythologized as having played an important part in postcolonial India's march toward self-sufficiency in food. Amul soon became the model for a national dairy development program and is thus credited with having initiated the "White Revolution" in India, the national dairy development program that made India the largest producer of milk in the world. On Amul, see Basu 2009.

28. Uttarakhand Tehsilwise and Districtwise Livestock Census 2012.

29. V. Padmakumar, C. T. Chacko, and Keith R. Sones "Improving Access to Breeding and Animal Health Services in Disadvantaged Locations," ILRI Research Brief 12, March 2014, http://www.himmotthan.in/UserFiles/files/improving_access_to_breeding_and_animal_health _research_5_3_15.pdf

30. "'Pahari' Cow Milk Cures Diseases, Claims Prof," *Decan Herald*, January 22, 2012, http:// www.deccanherald.com/content/221385/pahari-cow-milk-cures-diseases.html.

31. *Gau-kathas* appear to be a relatively recent phenomenon. They follow the pattern of Bhagvat-*kathas*, public readings and performances of Sanskrit texts, except that in this case the passages recited relate to cows.

32. Indeed, in response to fears that the indigenous cow is in danger of dying out because of aggressive interbreeding programs, efforts to protect and preserve the indigenous cow are underway. In 2014, a few months after coming to power, the national BJP government announced the "Rashtriya Gokul Mission," a program to preserve and protect indigenous breeds of cattle. Following suit, in 2015 the BJP government in the state of Haryana announced that they would no longer crossbreed indigenous breeds of cattle with foreign breeds. A year later, a minister in the government of Haryana caused an uproar when he called for the castration of foreign bulls on the grounds that their moral character was as suspect as the people from the countries they originally came from.

33. On the consumption of milk and kinship, see Carsten 1995.

34. Dean Nelson, "Drinking Milk from Non-Indian Cows 'Could Make Children Turn to Crime.'" *Telegraph* (London), April 24, 2015, http://www.telegraph.co.uk/news/worldnews/asia /india/11561612/Drinking-milk-from-non-Indian-cows-could-make-children-turn-to-crime .html.

35. As Charu Gupta (2001) points out, milk plays an important role in these imaginaries of

the cow as mother. "Like the mother's breasts, cows' udders [are] a metonym for nourishment and livelihood."

36. It was the recognition that the care of old cows was a difficult expense for rural and urban households to bear that moved the Supreme Court to rule against a national ban on cow slaughter in 1958. In 2005, "the court reversed a long line of decisions from 1958, accepting the assertion of the Gujarat government that new developments in technology and veterinary care had extended the economic life of the cow. Since 2008, the Supreme Court has retreated from the *Mirzapur* judgement, making it clear that an absolute ban on cow slaughter was not constitutionally required or desirable" (De 2013).

37. In 2017 the BJP won state elections in both Uttar Pradesh and Uttarakhand. Both governments immediately proclaimed that they would put an immediate end to cattle smuggling and illegal slaughter.

38. Buffaloes, too, are now being drawn into the broad category of "cow" that Hindu nationalists seek to protect. The legal scholar Madhav Khosla argues that this is the case because using a narrow legal definition of "cattle" in legislation that bans cow slaughter will "expose the religious motivation of the measure." "What's at Steak: The Constitutional Cost of Beef Ban," *Open Magazine*, June 16, 2017, http://www.openthemagazine.com/article/cover-story/what-s-at-steak-the-constitutional-cost-of-beef-ban.

39. For a powerful account of how kinship with the gods is established through such forms of exchange, see Ramberg 2014.

40. Even though the Jersey cow was not considered a ritually powerful being, there were some life rituals associated with cows that were too critical not to perform even in her case. For instance, as with *pahari* calves, there was always a naming ceremony for Jersey calves, held eleven days after their birth. And while people did sell milk to others who were not part of the household well before the prescribed twenty-two days, they would observe at least a couple of days of the prohibition. However, people's modification of these rituals at will was ascribed to the altered materiality of the Jersey and her significantly reduced ritual demands.

Chapter Four

1. This accusation was also picked up by the national press. Nihi Sharma Sahani, "Uttarakhand Monkeys May Have TB, Could be Threat to Humans: Experts," *Hindustan Times*, January 13, 2016, http://www.hindustantimes.com/india/uttarakhand-monkeys-may-have-tb-could-be-threat-to-humans-experts/story-dHXMVIqNUsUmpzvMeRQ38J.html.

2. On the complex relationship between humans and macaques, see Knight 2003; Fuentes and Wolfe 2002; Gumert, Fuentes, and Jones-Engel 2011; Radhakrishna, Huffman, and Sinha 2012.

3. People often claimed that they tolerated thefts by monkeys because of their similarity to humans. As Agustín Fuentes points out, no other creature on this planet has as much in common with humans—hands that can hold things, infant dependency, large brains, and behaviorally dynamic social complexity (Fuentes 2006).

4. On monkeys and folklore in South Asia, see Lutgendorf 2007; Pandian 2009; and Selby 2011.

5. John Hawley (1983, 8) notes that in some popular tellings of Krishna's adventures as a butter thief, he is said to "feed the leftovers to unruly, meddlesome monkeys." His observation that Krishna's thievery is widely interpreted as an act of love could explain why comparing Krishna's mischief to that of monkeys renders the latter's offences forgivable and even, dare I say, lovable.

6. Wendy Doniger (2009, 239) argues that in the Ramayana, monkeys "are the side shadows" of Rama and Lakshman. "They suggest what might have been. . . . They are, rather, parallel lives. . . . When Rama's cultural role as the perfect son and half brother prevents from expressing his personal resentment of his father and half brother, the monkeys do it for him."

7. As Dario Maestripieri (2007, 11) points out, rhesus macaques have been successful at displacing other macaque species from their native habitats as well as "in colonizing new habitats where no other macaques had gone before."

8. She was, in all probability, referring to an adult rhesus male who had been kicked out of the troop after reaching sexual maturity.

9. On fears about corporate control over natural resources, see Mathur 2015.

10. On the shifting population dynamics of rhesus monkeys in India, see Malik, Seth, and Southwick 1984; Malik 1992; and Southwick and Siddiqi 1988.

11. See also Sivaramakrishnan 2011 and Besky 2017.

12. For a close account of the everyday politics of the dynamics of monkey capture and control in Delhi, see Gandhi 2012.

13. See, e.g., Imam, Yahya, and Malik 2002 and Malik and Johnson 1994.

14. Faisal Mohammad Ali, "Row over Dehli's Errant Monkeys," BBC News, February 4, 2005, http://news.bbc.co.uk/2/hi/south_asia/4235811.stm.

15. J. Venkatesan, "Supreme Court Shifts Monkey Business to Delhi High Court," *Hindu*, February 15, 2007, http://www.thehindu.com/todays-paper/supreme-court-shifts-monkey-business-to-delhi-high-court/article1797516.ece.

16. Sajjad Hussain, "Why the Indian Parliament's Monkey Problem Has No Easy Solution," *Scroll.in*, August 1, 2014, http://scroll.in/article/672459/why-the-indian-parliaments-monkey-problem-has-no-easy-solution.

17. While the official version of events was that all these monkeys were sent to Asola and therefore kept within Delhi, forest officials in Uttarakhand suggested that many of the captured monkeys were, in fact, sent to other states—including Uttar Pradesh, Uttarakhand, and Haryana—because Asola could not contain such a large population. A couple of forest officials pointed to the accusation made by Maneka Gandhi that monkeys had starved to death in Asola because of corruption in the forest department and argued that the only flaw in her argument was that the monkeys had not died but had already been transported elsewhere instead. See notes 18 and 19 for more detail.

18. In 2015, when speaking with a news outlet about Gandhi's accusation that monkeys were being starved in Asola, the chief conservator of forests, Delhi, said "we have over 17,000 monkeys in this sanctuary and their number is continuously rising. *Neighboring states like Uttarakhand have already closed their doors for monkeys. In this situation, it's getting tough to accommodate such large number*" (emphasis mine). A retired forest official in Uttarakhand told me the very fact that the chief conservator of forests, Delhi, had mentioned Uttarakhand was an implicit admission that monkeys had been sent from Delhi to the mountains. "Why mention Uttarakhand at all?" he asked. "The official claim is that they have not sent any monkeys out of Delhi after 2007. The question the reporter asked was about Asola. I'll tell you why he mentioned it. It's because they have been sending monkeys here on the quiet [*chup-chaap*] for years now despite our government's saying no, and people are finding out. There has been no monkey census in Asola, so who knows how many monkeys are actually there in the sanctuary and how many have been dumped elsewhere."

19. Many agree that Asola has turned out to be far from ideal as a refuge for relocated monkeys. In 2014, Member of Parliament Maneka Gandhi called for an inquiry into the Delhi Forest

Department's management of Asola. She alleged that the money earmarked for feeding monkeys and planting fruit trees in the sanctuary was being stolen by officials. She added that almost a thousand monkeys had died of starvation in Asola because of the scam and that not a single tree had been planted despite claims to the contrary. The primatologist Iqbal Malik, who had previously advised the forest department on which trees and crops to plant in Asola, also accused the forest department of "unplanned and haphazard" management, saying that no vegetation had been planted or wells built in the sanctuary despite the fact that it was impossible to contain the monkeys in the sanctuary without making these arrangements. Darpan Singh, "Maneka Seeks Probe into Monkey Feed Expenses," *Hindustan Times* (Delhi), December 9, 2014, http://www.hindustantimes.com/india/maneka-seeks-probe-into-monkey-feed-expenses/story -MLccGfRCDewVNxPtJAYwhL.html.

20. Sharma Seema, "UP Dumping Its Monkeys in U'khand Secretly," *Times of India* (Delhi), December 11, 2015, http://epaperbeta.timesofindia.com/Article.aspx?eid=31808&articlexml=UP -dumping-its-monkeys-in-Ukhand-secretly-11122015017020

21. See note 1.

22. In Almora district, the Lok Prabandh Vikas Sansthan, an NGO that started out working on children's education, has now taken up the issue of translocated monkeys. "We had no choice," one of the NGO's employees told me as we sat in their small office in 2011, located just a few miles from the Binsar Wildlife Sanctuary in the middle of green fields. "All that the villagers in this area talk about now are the large numbers of monkeys who have invaded their homes and fields." When I asked him where the monkeys came from, he pointed in the general direction of Almora. "The municipal council of Almora brings monkeys from the town to the sanctuary," he said distractedly while shuffling through a sheaf of newspaper clippings to show me an article on a protest they had organized at the gates of the sanctuary in 2010. Handing me the piece he'd been looking for, he leaned back in his chair. Voice dripping with sarcasm, he asked, "You tell me, why would monkeys settle for that *vanvaas* (exile in the forest) when they can eat our food instead?"

23. For a wonderful essay on representations of rhesus macaques used in biomedical experiments in the United States between 1930 and 1960, see Ahuja 2012, 2016.

24. For more on fissioning, or "new group formation," see Malik, Seth, and Southwick 1985.

25. Chander Suta Dogra and Pramila N. Phatarphekar, "State in Monkey's Shadow," *Outlook*, September 27, 2004, http://www.outlookindia.com/magazine/story/state-in-monkeys-shadow /225254.

26. The D-Company is an organized crime syndicate founded and led by Dawood Ibrahim.

27. Bisht's views reminded me of Jason de León's (2015) observations about the role that nonhumans play in the policy of Prevention through Deterrence that Border Patrol has adopted to deal with migrants who seek to cross into the United States through its southern borderlands. De León (2015, 60) notes that while many "conceptualize the injuries and deaths that result from border crosser engagements with animals, terrain, and temperature as 'natural outcomes, . . . migrants have been purposefully funneled into the desert through various enforcement practices, a tactic that has enabled Border Patrol to outsource the work of punishment to actants such as mountains and extreme temperatures." Similarly, for Bisht, it was clear that the land mafia had outsourced its work to monkeys, who were running riot across the landscape and successfully clearing it of people. While Bisht had to admit that the monkeys did not *intend* to do this work for the land mafia, it was also undeniable, as far as he was concerned, that the land mafia could not have done their work without the monkeys.

28. Such stories of theft by monkeys in the city of Mathura abounded. In 2014, residents

of Mathura warned the president of India to watch out for the monkeys lest they attempted to take his glasses as well. Ishita Mishr, "Mathura Scare: Take Off Your Glasses, Mr President, or the Monkeys Will," *Times of India*, Novembeer 13, 2014, http://timesofindia.indiatimes.com /india/Mathura-scare-Take-off-your-glasses-Mr-President-or-the-monkeys-will/articleshow /45129598.cms.

29. As Jana Fortier (2009) notes in her ethnography of the Raute ethnic group in Nepal, the reluctance of Hindus to kill monkeys can be navigated in interesting ways. In western Nepal, the Raute are encouraged by Hindus to hunt monkeys on their land. This allows Hindus to get rid of the monkeys without having to kill them themselves.

30. Gandhi, Maneka, "Busting the 'Menace' Myth about Animals Calls for a Scientific Approach," *Firstpost*, April 4, 2016, http://www.firstpost.com/india/the-maneka-gandhi-column -busting-the-menace-myth-about-animals-calls-for-a-scientific-approach-2712166.html.

31. See, e.g., the article in *Down to Earth*, a fortnightly on the politics of environment and development: Shreeshan Venkatesh, "Why This Abandoned Village Is a Threat to Uttarakhand," *Down to Earth*, December 31, 2015, http://www.downtoearth.org.in/coverage/why-this -abandoned-village-is-a-threat-to-uttarakhand-52154.

32. Sometimes villagers address this shortage by hiring seasonal labor from Nepal and other parts of India.

33. Michael Levien (2015, 151) argues that the liberalization of India's economy after 1991 "unleashed increasing private demand for land for industry, infrastructure, and real estate. . . . In the 1990s the government loosened restrictions on bank lending to private developers. . . . This credit expansion, combined with growing demand for housing and office space, precipitated a dramatic real-estate boom by the mid-2000s."

34. Even though there is a legal ceiling on how much land outsiders can buy—1.25 nalis (20 nalis form an acre) per nuclear family—there are some situations in which they can acquire more than the allowance, e.g., outsiders can get special permission from the district magistrate to acquire a plot of land above the ceiling for agricultural purposes.

35. People often used the term *kothi* (bungalow) to distinguish the houses of outsiders from their own *ghar* (home).

36. The primatologist M. K. Neville (1968, 117), who studied troops of rhesus monkeys in Haldwani town and in the forest zone between 4,000 and 6,000 feet, also noted that forest monkeys in the Himalayas were shy and retiring, whereas the urban monkeys had lost their shyness in learning "how to utilize new food sources and by changing their ecological behavior."

37. See, e.g., Ciani 1986; Hrdy 1977; Singh 1968; Sengupta, McConkey, and Radhakrishna 2015; Southwick and Siddiqui 1994; and Teas et al 1982.

38. Farmers are often instructed by state officials and conservationists to plant crops like ginger that rhesus tend not to eat. Monkeys' dislike for ginger is captured in the Hindi proverb *bandar kya jaane adrak ka swad* (what does a monkey know of the taste of ginger). However, it appears from reports that monkeys in some parts have now acquired a taste for ginger. See, Anupam Chakravartty, "Out of Control: Why Monkeys Are a Menace," *Down to Earth*, August 31, 2015, http://www.downtoearth.org.in/coverage/out-of-control-why-monkeys-are-a-menace-50817.

39. It is certainly possible that the behavior of forest rhesus might change as an outcome of watching and learning from these commensal monkeys. Anindya Sinha (2003, 1029) points out that monkeys are able to comprehend each other's mental representation of the world, which allows them to "acquire knowledge of different attributes of [other] individuals, thus aiding their own decision making during social interactions."

Chapter Five

1. The Government of India was the official title of the central executive and legislative bodies for all of British India. The local governments were represented by the eight different provinces that made up the entity called British India, the part of India that was ruled directly by the British crown and did not include territory ruled by the native princes.

2. Letter from G. Bower, No. 1940, dated July 21, 1904, Uttar Pradesh State Archives, Forests, List No. 64, 1904–1925, Box No. 12, File No. 99, May-December 1904, Proceedings No. 7, Serial No. 6, p. 23.

3. Ibid.

4. Eventually, in light of opposition to the revised bill, the GOI abandoned plans for national legislation and asked provincial governments to issue their own regulations regarding the protection of wild animals in accordance with local regulations.

5. This was especially true in the wake of the Indian Mutiny of 1857, when colonial attitudes toward the natives hardened, and the need to establish racial difference between colonizer and colonized took on increased urgency (Pandian 2001).

6. On the relationship between *shikaris* and colonial hunters, see Hussain 2015; Pandian 1995; Pandian 2001; Rangarajan and Ranthambhore Foundation 2001; and Shreshth 2009.

7. See Skaria 2002 on the politics of wildness in colonial and postcolonial India.

8. Quoted in Shreshth 2009, 350.

9. Indeed, repeated, low-intensity crop raiding by smaller species, especially ruminants and herbivores, has proven to be more economically damaging over time than high-intensity conflict with large terrestrial mammals (Ogra 2009).

10. Mayank Aggarwal, "Wild Pigs Declared Vermin in Uttarakhand for a Year," *Livemint*, February 10, 2016http://www.livemint.com/Politics/TSea6JoX9dNMS2eibTMW9M/Wild-pigs -declared-vermin-in-Uttarakhand-for-a-year.html.

11. Seema Sharma, "Wild Boar Gets a Vermin Tag, Wildlife Activists Up in Arms," *Times of India*. February 10, 2016, http://timesofindia.indiatimes.com/city/dehradun/Wild-boar-gets -vermin-tag-wildlife-activists-up-in-arms/articleshow/50935650.cms.

12. Dinesh is not the only person who claims that wild boar in Uttarakhand do not like the sound of Punjabi music. In December 2015, it was reported that residents of a village in Nainital district, not far from where I conducted fieldwork, were "blasting" songs by the rapper Yo Yo Honey Singh from a loudspeaker in order to scare wild boar away from their fields. Ashmita Nayar, "Farmers Are Blaring Yo Yo Honey Singh's Music to Scare Away Wild Animals and It's Working," *Huffington Post* (India), February 12, 2016, http://www.huffingtonpost.in/2015/12/02 /honey-singh-uttarkhand-lo_n_8695396.html.

13. See, e.g., Beinart 2015 and Knight 2003.

14. On the complicated relationships entailed by domestication, see Cassidy and Mullin 2007.

15. On disgust as a means of marking difference, see Ghassem Fachandi 2010; Guru 2009; and Pinto 2006

16. The reasons for Chandan Da's silence were complex. Chandan Da was a Shilpakar from what he described as a *thul jati*, that is, a *jati* that was considered at the top of the hierarchy among different Shilpakar *jatis*. His family were Tamtas, coppersmiths, and he considered himself superior to *halis* (literally, ploughmen), Dalits who were forced to work as agrarian labor for upper castes. While he resented Manish's othering of lower castes, I knew he, too, considered Balmikis unclean and ritually polluting. He had once told me about the time he had rushed home to sprinkle cow urine on himself in an effort to purify his body after accepting water from a man

in Almora. The man, he found out only after drinking the offered glass of water, was a Balmiki. "They should not hire such people in restaurants," he told me indignantly. "This is why I usually never eat or drink anything outside." As Michael Moffatt (1979) points out, "untouchables are not detached from the 'rationalisation' of the system. . . . They recreate among themselves the entire set of institutions and of ranked relations from which they have been excluded by the higher castes by reason of their extreme lowness." Indeed, the Balmikis are a group who are frequently discriminated against, even by other low-caste groups who place themselves higher in the caste hierarchy vis-à-vis the former. On Balmikis and their relationship to Hinduism, see Lee 2015.

17. On the romance of wilderness and its hidden exclusions, see Cronon 1996; Jacoby 2001; Neumann 1998.

Chapter Six

1. Also see Flueckiger 1996; Gold 1997; Raheja and Gold 1994; Narayan and Sood 1997; Ramberg 2014; Trawick 1988; and Wadley 1994.

2. On bear ceremonialism, see Hallowell 1926 and Ohnuki-Tierney 1974.

3. See as one example "A Thief, A Ram, A Bear, and a Horse" in Ramanujan, Blackburn, and Dundes 1997.

4. See Messner 2000.

5. In this genre, you could go to bed with a human or deity and wake up the next morning to find that who you really went to bed with last night was an animal. The reverse is also possible: you could go to bed with an animal and wake up to find that the being you bedded is, in reality, a human or a god. For more on such animal tales, see Blackburn 1995; Doniger 2000; and Karnad 1990.

6. I am indebted to Naisargi Dave for pointing out that these stories constitute a genre.

7. Ann Gold (Raheja and Gold 1994, 133–34) finds a similar disjuncture between the public and private behavior of husbands as portrayed in women's songs in Rajasthan. She writes that "while men appear in some songs as publicly defying patrilineal dicta, they are portrayed in others as feigning, before an audience, adherence to norms concerning the priority to be placed on patrilineal solidarities rather than on intimacy and solidarity with the wife, while at the same time privately undermining them."

8. In her work on the forms of resistance enacted by Bedouin women, Lila Abu-Lughod (1990, 45) notes that women's sexual irreverence reveals how "the code of sexual morality and the ideology of sexual difference are forms of men's power. Women seem only too glad when men fail to live up to the ideals of autonomy and manhood, the ideals on which their alleged moral superiority and social precedence are based, especially if they fail as a result of sexual desire."

9. Gail Omvedt similarly observes that "caste can survive only if women's sexuality is controlled." Meena Kandasamy, "In Conversation: Gail Omvedt," *UltravViolet* (blog), February 29, 2008, https://youngfeminists.wordpress.com/2008/02/29/120/.

10. On the complex intersection of expectations around fertility and restraint, see Raheja and Gold 1994.

11. On discourses and practices of shame and modesty in South Asia, see Raheja and Gold 1994. For an examination of ideas about modesty in a different context, see Abu-Lughod 1986.

12. For a powerful critique of the colonial origins of the "Third World Woman" who is represented in Western feminist scholarship as traditional and oppressed, see Chandra Mohanty 1988.

13. See, e.g., Roychowdhury 2013.

14. See Charoo et al 2011; Sathyakumar 2001.

15. Kinloch 1885, 48–49.

16. On colonial programs of vermin eradication, see Pandian 2001; Rangarajan 1998; Rangarajan and Ranthambhore Foundation 2001; Sivaramakrishnan 1999.

17. Uttar Pradesh State Archives, Forests, 1912 List No. 64, 1904–1925, Box No. 125, File No. 32, A Proceedings Numbers 7–10, Serial Numbers 1–4.

18. Letter from B. B. Osmaston, Esq. Conservator of Forests, Western Circle to Chief Secretary to Government, United Provinces, March 11, 1912, p. 3, Uttar Pradesh State Archives, Forests, 1912 List No. 64, 1904–1925, Box No. 125, File No. 32, A Proceedings Numbers 7–10, Serial Number 1.

19. Ibid.

20. Letter from Chief Secretary to Government, United Provinces, to Conservator of Forests, Western Circle, dated June 11, 1912, Uttar Pradesh State Archives, Forests, 1912 List No. 64, 1904–1925, Box No. 125, File No. 32, A Proceedings Numbers 7–10, Serial Number 2.

21. See Saberwal, Rangarajan, and Kothari 2001, 39.

22. Sometimes, however, bears make spectacular appearances. In 2016, a video of a bear who, after walking into a hotel in Nainital and attacking the owner, went for a leisurely swim in the lake before disappearing into the forest, went viral on Facebook and WhatsApp. Most people who shared the video had never seen a bear in person before and were shocked that their first encounter with one had been in the unexpected space of a city. People's perceptions about the unpredictability of bears were only reinforced by this encounter.

Epilogue

1. A team of wildlife biologists led by Vidya Athreya (Athreya et al. 2014) analyzed eighty-five samples of leopard scat and found that 87 percent of prey biomass was made up of domestic animals, of which dogs constituted 39 percent. They argue that "the standing biomass of dogs and cats alone was sufficient to sustain the high density of carnivores at the study site."

2. I am grateful to Naisargi Dave for posing this question to me.

3. Heather Davis and Paige Sarlin, "'On the Risk of a New Relationality': An Interview with Lauren Berlant and Michael Hardt," in "Remaking the Commons," ed. Matthew MacClellan and Margrit Talpularu, special issue, *Reviews in Cultural Theory* 2, no. 3 (2011), http://reviewsinculture.com/2012/10/15/on-the-risk-of-a-new-relationality-an-interview-with-lauren-berlant-and-michael-hardt/.

4. See, e.g., Howe and Pandian 2017 and Bauer and Bhan 2016.

References

Abu-Lughod, Lila. 1986. *Veiled Sentiments: Honor and Poetry in a Bedouin Society*. Berkeley: University of California Press.

———. 1990. "The Romance of Resistance: Tracing Transformations of Power through Bedouin Women." *American Ethnologist* 17 (1): 41–55.

Adcock, C. S. 2010. "Sacred Cows and Secular History: Cow Protection Debates in Colonial North India." *Comparative Studies of South Asia, Africa and the Middle East* 30 (2): 297–311.

Agarwal, Bina. 1992. "The Gender and Environment Debate: Lessons from India." *Feminist Studies* 18 (1): 119.

———. 1994. "Gender, Resistance and Land: Interlinked Struggles over Resources and Meanings in South Asia." *Journal of Peasant Studies* 22 (1): 81–125.

Ahmed, Sara. 2010. *The Promise of Happiness*. Durham NC: Duke University Press.

Ahuja, Neel. 2012. "Macaques and Biomedicine: Notes on Decolonization, Polio and Changing Representations in Indian Rhesus in the United States, 1930–1960." In *The Macaque Connection: Cooperation and Conflict between Humans and Macaques*, edited by Sindhu Radhakarishna, Michael A. Huffman, and Anindya Sinha, 71–99." New York: Springer.

———. 2016. *Bioinsecurities: Disease Interventions, Empire, and the Government of Species*. Durham, NC: Duke University Press.

Alaimo, Stacy. 2010. *Bodily Natures: Science, Environment, and the Material Self*. Bloomington: Indiana University Press.

———. 2016. *Exposed: Environmental Politics and Pleasures in Posthuman Times*. Minneapolis: University of Minnesota Press.

Allen, Jafari Sinclaire, and Ryan Cecil Jobson. 2016. "The Decolonizing Generation: (Race and) Theory in Anthropology since the Eighties." *Current Anthropology* 57, no. 2 (April): 129–48.

Al-Mohammad, Hayder, and Daniela Peluso. 2012. "Ethics and the 'Rough Ground' of the Everyday: The Overlappings of Life in Postinvasion Iraq." *HAU: Journal of Ethnographic Theory* 2 (2): 42–58.

Amin, Shahid. 2016. *Conquest and Community: The Afterlife of Warrior Saint Ghazi Miyan*. Chicago: University of Chicago Press.

Anderson, Virginia DeJohn. 2004. *Creatures of Empire: How Domestic Animals Transformed Early America*. Oxford: Oxford University Press.

Archambault, Julie Soleil. 2016. "Taking Love Seriously in Human-Plant Relations in Mozambique: Toward an Anthropology of Affective Encounters." *Cultural Anthropology* 31 (2): 244–71.

Aryal, M. 1994. "Axing Chipko." *Himal* 7 (1): 8–23.

Asad, Talal. 1993. *Genealogies of Religion: Discipline and Reasons of Power in Christianity and Islam*. Baltimore, MD: Johns Hopkins University Press.

Athreya, Vidya, Morten Odden, John D. C. Linnell, Jagdish Krishnaswamy, and K. Ullas Karanth. 2014. "A Cat among the Dogs: Leopard *Panthera pardus* Diet in a Human-Dominated Landscape in Western Maharashtra, India." *Oryx* 50 (1): 156–62.

Atkinson, Edwin T. 1973. *The Himalayan Gazetteer*. Delhi: Cosmo.

Bacchetta, Paola. 1999. "When the (Hindu) Nation Exiles Its Queers." *Social Text* (61): 141–66.

Barad, Karen. 2012. "Interview." In *New Materialism: Interviews and Cartographies*, edited by Rick Dolphijn and Iris van der Tuin, 48–70. Ann Arbor, MI: Open Humanities Press.

Basu, Pratyusha. 2009. *Villages, Women, and the Success of Dairy Cooperatives in India: Making Place for Rural Development*. Amherst, NY: Cambria Press.

Bauer, Andrew M., and Mona Bhan. 2016. "Welfare and the Politics and Historicity of the Anthropocene." *South Atlantic Quarterly* 115 (1): 61–87.

Bayly, C. A. 1988. *Indian Society and the Making of the British Empire*. Cambridge: Cambridge University Press.

Baynes-Rock, Marcus. 2015. *Among the Bone-Eaters: Encounters with Hyenas in Harar*. University Park: Pennsylvania State University Press.

Beinart, William. 2015. "Reflecting on Unruliness." In *Unruly Environments*, edited by Siddhartha Krishnan, Christopher L. Pastore, and Samuel Temple, 69–74. Munich: Rachel Carson Center.

Bennett, Jane. 2010. *Vibrant Matter: A Political Ecology of Things*. Durham, NC: Duke University Press.

Berlant, Lauren. 2011a. *Cruel Optimism*. Durham, NC: Duke University Press.

———. 2011b. "A Properly Political Concept of Love: Three Approaches in Ten Pages." *Cultural Anthropology* 26 (4): 683–91.

Berreman, Gerald Duane. 1963. *Hindus of the Himalayas*. Berkeley: University of California Press.

Berry, Kim, and Shubhra Gururani. 2014. "Special Section: Gender in the Himalaya." *Himalaya, the Journal of the Association for Nepal and Himalayan Studies* 34 (1): art. 9.

Besky, Sarah. 2013. *The Darjeeling Distinction Labor and Justice on Fair-Trade Tea Plantations in India*. Berkeley: University of California Press.

———. 2017. "The Land in Gorkhaland: On the Edges of Belonging in Darjeeling, India." *Environmental Humanities* 9 (1): 18–39.

Bessire, Lucas. 2014. *Behold the Black Caiman: A Chronicle of Ayoreo Life*. Chicago: University of Chicago Press.

Biardeau, Madeleine, Alf Hiltebeitel, Marie Louise Reiniche, and Taï Walker. 2004. *Stories about Posts: Vedic Variations around the Hindu Goddess*. Chicago: University of Chicago Press.

Bieder, Robert E. 2005. *Bear*. London: Reaktion Books.

Biehl, João. 2013. "Ethnography in the Way of Theory." *Cultural Anthropology* 28 (4): 573–97.

Bird Rose, Deborah. 2011. *Wild Dog Dreaming: Love and Extinction*. Charlottesville: University of Virginia Press.

Bishop, Naomi, Sarah Blaffer Hrdy, Jane Teas, and James Moore. 1981. "Measures of Human Influence in Habitats of South Asian Monkeys." *International Journal of Primatology* 2 (2): 153–67.

Blackburn, Stuart H., ed. 1995. "Coming Out of His Shell: Animal-Husband Tales from India." In

Syllables of Sky: Studies in South Indian Civilization, edited by David Shulman, 43–75. New Delhi: Oxford University Press.

Blackburn, Stuart H., and A. K. Ramanujan. 1986. *Another Harmony: New Essays on the Folklore of India*. Berkeley: University of California Press.

Blanchette, Alex. 2015. "Herding Species: Biosecurity, Posthuman Labor, and the American Industrial Pig." *Cultural Anthropology* 30 (4): 640–69.

Butler, Judith. 2004. *Undoing Gender*. New York: Routledge. doi:9780415969239.

Cadena, Marisol de la. 2015. *Earth Beings: Ecologies of Practice across Andean Worlds*. Durham, NC: Duke University Press.

Campbell, Ben. 2013. *Living between Juniper and Palm: Nature, Culture, and Power in the Himalayas*. New Delhi: Oxford University Press.

Carrithers, Michael, Matei Candea, Karen Sykes, Martin Holbraad, and Soumhya Venkatesan. 2010. "Ontology Is Just Another Word for Culture: Motion Tabled at the 2008 Meeting of the Group for Debates in Anthropological Theory, University of Manchester." *Critique of Anthropology* 30 (2): 152–200.

Carsten, Janet. 1995. "The Substance of Kinship and the Heat of the Hearth: Feeding, Personhood, and Relatedness among Malays in Pulau Langkawi." *American Ethnologist* 22 (2): 223–41.

———. 2000. *Cultures of Relatedness: New Approaches to the Study of Kinship*. Cambridge: Cambridge University Press.

———. 2013. "What Kinship Does—and How." *HAU: Journal of Ethnographic Theory* 3 (2). https://www.haujournal.org/index.php/hau/article/view/hau3.2.013/376.

Cassidy, Rebecca. 2002. *The Sport of Kings: Kinship, Class, and Thoroughbred Breeding in Newmarket*. Cambridge: Cambridge University Press.

Cassidy, Rebecca, and Molly H. Mullin. 2007. *Where the Wild Things Are Now: Domestication Reconsidered*. Oxford: Berg.

Chakravarti, Uma. 1993. "Conceptualising Brahmanical Patriarchy in Early India: Gender, Caste, Class and State." *Economic and Political Weekly* 28 (14): 579–85.

Charoo, Samina A., Lalit K. Sharma, and S. Sathyakumar. 2011. "Asiatic Black Bear–Human Interactions around Dachigam National Park, Kashmir, India." *Ursus* 22 (2): 106–13.

Chatterji, Joya. 1994. *Bengal Divided: Hindu Communalism and Partition, 1932–1947*. Cambridge: Cambridge University Press.

Chauhan, N. P. S. 2011. "Human Casualties and Agricultural Crop Raiding by Wild Pigs and Mitigation Strategies in India." *Julius Kühn-Archiv* 432: 192–92.

Chen, Mel Y. 2012. *Animacies: Biopolitics, Racial Mattering, and Queer Affect*. Perverse Modernities. Durham, NC: Duke University Press.

Chigateri, Shraddha. 2011. "Negotiating the 'Sacred' Cow: Cow Slaughter and the Regulation of Difference in India." In *Democracy, Religious Pluralism and the Liberal Dilemma of Accommodation*, edited by Monica Mookherjee, 137–59. Dordrecht, The Netherlands: Springer.

Choy, Timothy K. 2011. *Ecologies of Comparison: An Ethnography of Endangerment in Hong Kong*. Durham, NC: Duke University Press.

Ciani, Andrea Camperio. 1986. "Intertroop Agonistic Behavior of a Feral Rhesus Macaque Troop Ranging in Town and Forest Areas in India." *Aggressive Behavior* 12 (6): 433–39.

Coetzee, J. M., Marjorie Gerber, Peter Singer, Wendy Doniger, and Barbara Smuts. *The Lives of Animals*. Edited by Amy Gutmann. Princeton, NJ: Princeton University Press.

Comaroff, Jean, and John L. Comaroff. 2001. "Naturing the Nation: Aliens, Apocalypse, and the Postcolonial State." *Social Identities* 7 (2): 233–65.

Coole, Diana H., and Samantha Frost. 2010. *New Materialisms: Ontology, Agency, and Politics.* Durham NC: Duke University Press.

Cormier, Loretta A. 2003. *Kinship with Monkeys: The Guajá Foragers of Eastern Amazonia.* New York: Columbia University Press.

Cronon, William. 1996. "The Trouble with Wilderness; or, Getting Back to the Wrong Nature." *Environmental History* 1 (1): 7–28.

Daniel, E. Valentine. 1984. *Fluid Signs: Being a Person the Tamil Way.* Berkeley: University of California Press.

Das, Veena. 1983. "Language of Sacrifice." *Man* 18 (3): 445–62.

———. 2007. *Life and Words: Violence and the Descent into the Ordinary.* Berkeley: University of California Press.

———. 2012. "Ordinary Ethics." In *A Companion to Moral Anthropology*, edited by Didier Fassin, 133–49: Malden, MA: Wiley-Blackwell.

———. 2013. "Being Together with Animals: Death, Violence and Noncruelty in Hindu Imagination." In *Living Beings: Perspectives on Interspecies Engagements*, edited by Penelope Dransart, 1–16. London: Bloomsbury.

Dave, Naisargi N. 2012. *Queer Activism in India: A Story in the Anthropology of Ethics.* Durham, NC: Duke University Press.

———. 2014. "Witness: Humans, Animals and the Politics of Becoming." *Cultural Anthropology* 29 (3): 433–56.

———. 2015. "Love and Other Injustices: On Indifference to Difference." Humanities Futures, Franklin Humanities Institute, Duke University. https://humanitiesfutures.org/papers/845/.

De, Rohit. "Who Moved My Beef?" *Hindu*, Business Line, November 18, 2013, http://www.the hindubusinessline.com/opinion/who-moved-my-beef/article5364664.ece.

De León, Jason. 2015. *The Land of Open Graves: Living and Dying on the Migrant Trail.* Berkeley: University of California Press.

Desjarlais, Robert R. 2016. *Subject to Death: Life and Loss in a Buddhist World.* Chicago: University of Chicago Press.

Desmond, Jane C. 2016. *Displaying Death and Animating Life: Human-Animal Relations in Art, Science, and Everyday Life.* Chicago: University of Chicago Press.

Despret, Vinciane. 2008. "The Becomings of Subjectivity in Animal Worlds." *Subjectivities* 23: 123–39.

———. 2013. "Responding Bodies and Partial Affinities in Human-Animal Worlds." *Theory, Culture & Society* 30 (7/8): 51–76.

De Waal, Frans. 1997. "Are We in Anthropodenial?" *Discover* 18 (7): 50–53.

———. 2016. *Are We Smart Enough to Know How Smart Animals Are?* New York: W. W. Norton.

Dhavamony, Mariasusai. 1973. *Phenomenology of Religion.* Rome: Gregorian University Press.

Doniger, Wendy. 1999. *Splitting the Difference: Gender and Myth in Ancient Greece and India.* Chicago: University of Chicago Press.

———. 2000. *The Bedtrick: Tales of Sex and Masquerade.* Chicago: University of Chicago Press.

———. 2009. *The Hindus: An Alternative History.* New York: Penguin.

Douglas, Mary. 1966. *Purity and Danger: An Analysis of Concepts of Pollution and Taboo.* New York: Praeger.

Duffy, Mignon. 2007. "Doing the Dirty Work." *Gender & Society* 21 (3): 313–36.

Durkheim, Émile. 2001 *The Elementary Forms of Religious Life.* Translated by Carol Cosman. Abridged with an introduction and notes by Mark S. Cladis. Oxford: Oxford University Press.

Dyson, Jane. 2014. *Working Childhoods: Youth, Agency and the Environment in India*. Cambridge: Cambridge University Press.

Edelman, Lee. 2004. *No Future: Queer Theory and the Death Drive*. Durham, NC: Duke University Press.

Engineer, Asghar Ali. 1995. *Communalism in India: A Historical and Empirical Study*. New Delhi: Vikas.

Escobar, Arturo, and Sonia E. Alvarez. 1992. *The Making of Social Movements in Latin America: Identity, Strategy, and Democracy*. Boulder, CO: Westview.

Evans-Pritchard, E. E. 1954. "The Meaning of Sacrifice among the Nuer." *Journal of the Royal Anthropological Institute of Great Britain and Ireland* 84 (1/2): 21–33.

———. 1956. *Nuer Religion*. Oxford: Clarendon.

Ewing, Katherine P. 1994. "Dreams from a Saint: Anthropological Atheism and the Temptation to Believe." *American Anthropologist* 96 (3): 571–83.

Fanon, Frantz. 1963. *The Wretched of the Earth*. New York: Grove.

Faubion, James D. 2011. *An Anthropology of Ethics*. Cambridge: Cambridge University Press. doi:40019446823.

Flueckiger, Joyce Burkhalter. 1996. *Gender and Genre in the Folklore of Middle India*. Ithaca, NY: Cornell University Press.

Fortier, Jana. 2009. *Kings of the Forest: The Cultural Resilience of Himalayan Hunter-Gatherers*. Honolulu: University of Hawai'i Press.

Franklin, Sarah, and Susan McKinnon. 2001. *Relative Values: Reconfiguring Kinship Studies*. Durham, NC: Duke University Press.

Freitag, Sandria B. 1980. "Sacred Symbol as Mobilizing Ideology: The North Indian Search for a 'Hindu' Community." *Comparative Studies in Society and History* 22 (4): 597–625.

Freud, Sigmund. 1918. *Totem and Taboo: Resemblances between the Psychic Lives of Savages and Neurotics*. Translated by A. A. Brill. New York: Moffat, Yard.

Fuentes, Agustín. 2006. "Human-Nonhuman Primate Interconnections and Their Relevance to Anthropology." *Ecological and Environmental Anthropology* 2 (2): 1–11.

Fuentes, Agustín. 2010. "Naturalcultural Encounters in Bali: Monkeys, Temples, Tourists, and Ethnoprimatology." *Cultural Anthropology* 25 (4): 600–24.

Fuentes, Agustín, and Linda D. Wolfe, eds. 2002. *Primates Face to Face the Conservation Implications of Human-Nonhuman Primate Interconnections*. Cambridge: Cambridge University Press.

Fuller, C. J. 1992. *The Camphor Flame: Popular Hinduism and Society in India*. Princeton, NJ: Princeton University Press.

Gandhi, Ajay. 2012. "Catch Me If You Can: Monkey Capture in Delhi." *Ethnography* 13 (1): 43–56.

Gane, Nicholas. 2006. "When We Have Never Been Human, What Is to Be Done?" *Theory, Culture & Society* 23 (7/8): 135–58.

Garcia, Maria Elena. Forthcoming. "Loving Guinea Pigs in Peru: Life, Death, and the (Im)possibilities of Collaborative Multispecies Ethnography". *HAU: Journal of Ethnographic Theory*.

Ghassem-Fachandi, Parvis. 2010. "On the Political Use of Disgust in Gujarat." *South Asian History and Culture* 1 (4): 557–76.

———. 2012. *Pogrom in Gujarat Hindu Nationalism and Anti-Muslim Violence in India*. Princeton, NJ: Princeton University Press.

Gillespie, Kathryn, and Rosemary-Claire Collard. 2015. *Critical Animal Geographies: Politics, Intersections, and Hierarchies in a Multispecies World*. Abingdon, Oxon: Routledge.

Girard, René. 1979. *Violence and the Sacred*. Baltimore, MD: Johns Hopkins University Press.

Gluck, John P. 2016. *Voracious Science and Vulnerable Animals: A Primate Scientist's Ethical Journey*. Chicago: University of Chicago Press.

Gold, Ann Grodzins. 1997. "Outspoken Women: Representations of Female Voices in a Rajasthani Folklore Community." *Oral Tradition* 12 (1): 103–33.

Gold, Ann Grodzins, and Bhoju Ram Gujar. 1997. "Wild Pigs and Kings: Remembered Landscapes in Rajasthan." *American Anthropologist* 99 (1): 70–84.

Gould, William. 2012. *Religion and Conflict in Modern South Asia*. New York: Cambridge University Press.

Govindrajan, Radhika. 2013. "Beastly Intimacies: Interspecies Relations in India's Central Himalayas." PhD diss., Yale University.

———. 2014. "How to Be Hindu in the Himalayas: Conflicts over Animal Sacrifice in Uttarakhand." In *Shifting Ground: People, Animals, and Mobility in India's Environmental History*, edited by K. and Mahesh Rangarajan Sivaramakrishnan, 204–27. Delhi: Oxford University Press.

———. 2015a. "The Goat That Died for Family": Animal Sacrifice and Interspecies Kinship in India's Central Himalayas." *American Ethnologist* 42 (3): 504–19.

———. 2015b. "The Man-Eater Sent by God: Unruly Interspecies Intimacy in India's Central Himalayas." In *Unruly Environments*, edited by Siddharta Krishnan, Christopher L. Pastore, and Samuel Temple, 33–38. Munich: Rachel Carson Center.

———. 2015c. "Monkey Business: Macaque Translocation and the Politics of Belonging in India's Central Himalayas." *Comparative Studies of South Asia, Africa and the Middle East* 35 (2): 246–62.

Graeber, David. 2015. "Radical Alterity Is Just Another Way of Saying 'Reality': A Reply to Eduardo Viveiros De Castro." *HAU: Journal of Ethnographic Theory* 5 (2): 41.

Guha, Ramachandra. 1990. *The Unquiet Woods: Ecological Change and Peasant Resistance in the Himalaya*. Berkeley: University of California Press.

Gupta, Charu. 2001. "The Icon of Mother in Late Colonial North India." *Economic and Political Weekly* 36 (45): 4291–99.

Gumert, Michael D., Agustín Fuentes, and Lisa Jones-Engel. 2011. *Monkeys on the Edge: Ecology and Management of Long-Tailed Macaques and Their Interface with Humans*. Cambridge: Cambridge University Press.

Guru, Gopal. 2009. *Humiliation: Claims and Context*. New Delhi: Oxford University Press.

Gururani, Shubhra. 2002. "Forests of Pleasure and Pain: Gendered Practices of Labor and Livelihood in the Forests of the Kumaon Himalayas, India." *Gender, Place & Culture* 9 (3): 229–43.

Halberstam, Jack. 2014. "Wildness, Loss, Death." *Social Text* 32 (4): 137–48.

Hallowell, A. Irving. 1926. "Bear Ceremonialism in the Northern Hemisphere." *American Anthropologist* 28 (1): 1–175.

Hansen, Thomas Blom. 1999. *The Saffron Wave: Democracy and Hindu Nationalism in Modern India*. Princeton, NJ: Princeton University Press.

Haraway, Donna. 1988. "Situated Knowledges: The Science Question in Feminism and the Privilege of Partial Perspective." *Feminist Studies* 14 (3): 575–99.

Haraway, Donna Jeanne. 1989. *Primate Visions: Gender, Race, and Nature in the World of Modern Science*. New York: Routledge.

———. 1991. *Simians, Cyborgs, and Women: The Reinvention of Nature*. New York: Routledge.

———. 2003. *The Companion Species Manifesto: Dogs, People, and Significant Otherness*. Chicago: Prickly Paradigm Press. doi:9780971757585.

————. 2008. *When Species Meet*. Minneapolis: University of Minnesota Press. doi:9780816650460.

————. 2016. *Staying with the Trouble: Making Kin in the Chthulucene*. Durham, NC: Duke University Press.

Harris, Marvin. 1966. "The Cultural Ecology of India's Sacred Cattle." *Current Anthropology* 7 (1): 51–66.

Hawley, John Stratton. 1983. *Krishna, the Butter Thief*. Princeton, NJ: Princeton University Press.

Hiltebeitel, Alf. 1980. "Rama and Gilgamesh: The Sacrifices of the Water Buffalo and the Bull of Heaven." *History of Religions* 19 (3): 187–223.

Howe, Cymene. 2015. "Porous Pleasures." Theorizing the Contemporary, *Cultural Anthropology* website, July 21. https://culanth.org/fieldsights/701-porous-pleasures.

Howe, Cymene, and Anand Pandian. 2016. "Introduction: Lexicon for an Anthropocene Yet Unseen." Theorizing the Contemporary, *Cultural Anthropology* website, January 21. https://culanth.org/fieldsights/788-introduction-lexicon-for-an-anthropocene-yet-unseen.

Hrdy, Sarah Blaffer. 1977. *The Langurs of Abu: Female and Male Strategies of Reproduction*. Cambridge, MA: Harvard University Press.

Hubert, Henri, and Marcel Mauss. 1964. *Sacrifice: Its Nature and Function*. Chicago: University of Chicago Press.

Hussain, Shafqat. 2015. *Remoteness and Modernity: Transformation and Continuity in Northern Pakistan*. New Haven, CT: Yale University Press.

Imam, Ekwal, H. S. A. Yahya, and Iqbal Malik. 2002. "A Successful Mass Translocation of Commensal Rhesus Monkeys *Macaca mulatta* in Vrindaban, India." *Oryx* 36 (1): 87–93.

Ingold, Tim. 2000. *The Perception of the Environment: Essays on Livelihood, Dwelling and Skill*. London: Routledge.

Jacoby, Karl. 2001. *Crimes against Nature: Squatters, Poachers, Thieves, and the Hidden History of American Conservation*. Berkeley: University of California Press.

Jaffrelot, Christophe. 1996. *The Hindu Nationalist Movement in India*. New York: Columbia University Press.

————. 2007. *Hindu Nationalism: A Reader*. Princeton, NJ: Princeton University Press.

————. 2008. "Hindu Nationalism and the (Not So Easy) Art of Being Outraged: The Ram Setu Controversy." *South Asia Multidisciplinary Academic Journal* 2. https://samaj.revues.org/1372.

Jalais, Annu. 2010. *Forest of Tigers: People, Politics and Environment in the Sundarbans*. London: Routledge.

Jeffery, Laura, and Matei Candea. 2006. "The Politics of Victimhood." *History and Anthropology* 17 (4): 287–96.

Jodhka, Surinder S., and Murli Dhar. 2003. "Cow, Caste and Communal Politics: Dalit Killings in Jhajjar." *Economic and Political Weekly* 38 (3): 174–76.

Jones, Kenneth W. 1992. *Religious Controversy in British India: Dialogues in South Asian Languages*. SUNY Series in Religious Studies. Albany: State University of New York Press.

Joshi, Sanjay. 2015. "Juliet Got It Wrong: Conversion and the Politics of Naming in Kumaon, ca. 1850–1930." *Journal of Asian Studies* 74 (4): 843–62.

Karnad, Girish Raghunath. 1990. *Nāga-Mandala: Play with a Cobra*. Delhi: Oxford University Press.

Keenan, Dennis King. 2005. *The Question of Sacrifice: Studies in Continental Thought*. Bloomington: Indiana University Press.

Kim, Claire Jean. 2015. *Dangerous Crossings: Race, Species, and Nature in a Multicultural Age*. New York: Cambridge University Press.

King, Barbara. 2013. *How Animals Grieve.* Chicago: University of Chicago Press.

Kinloch, Alexander Angus Airlie. 1885. *Large Game Shooting in Thibet, the Himalayas, and Northern India.* Calcutta: W. Thacker.

Kirksey, S. Eben, and Stefan Helmreich. 2010. "The Emergence of Multispecies Ethnography." *Cultural Anthropology* 25 (4): 545–76.

Knight, John. 2003. *Waiting for Wolves in Japan: An Anthropological Study of People-Wildlife Relations.* Oxford: Oxford University Press.

Kohn, Eduardo. 2007. "How Dogs Dream: Amazonian Natures and the Politics of Transspecies Engagement." *American Ethnologist* 34 (1): 3–24.

———. 2013. *How Forests Think: Toward an Anthropology beyond the Human.* Berkeley: University of California Press.

———. 2015. "Anthropology of Ontologies." *Annual Review of Anthropology* 44 (1): 311–27.

Kosek, Jake. 2010. "Ecologies of Empire: On the New Uses of the Honeybee." *Cultural Anthropology* 25 (4): 650–78.

Kressler, Brad. 2010. *Goat Song: A Seasonal Life, A Short History of Herding, and the Art of Making Cheese.* New York: Scribner's.

Kuzniar, Alice A. 2006. *Melancholia's Dog.* Chicago: University of Chicago Press.

Laidlaw, James. 2014. *The Subject of Virtue: An Anthropology of Ethics and Freedom.* New Departures in Anthropology. New York: Cambridge University Press.

Lamb, Sarah. 2000. *White Saris and Sweet Mangoes: Aging, Gender, and Body in North India.* Berkeley: University of California Press.

Lambek, Michael. 2010. *Ordinary Ethics: Anthropology, Language, and Action.* New York: Fordham University Press.

Latour, Bruno. 2005. *Reassembling the Social: An Introduction to Actor-Network-Theory.* Oxford; New York: Oxford University Press.

———. 2010. "An Attempt at a 'Compositionist Manifesto.'" *New Literary History* 41 (3): 471–90.

Lee, Joel. 2015. "Jagdish, Son of Ahmad: Dalit Religion and Nominative Politics in Lucknow." *South Asia Multidisciplinary Academic Journal* 11. http://samaj.revues.org/3919.

Lempert, Michael. 2014. "Uneventful Ethics." *HAU: Journal of Ethnographic Theory* 4 (1): 465–72.

Levien, Michael. 2015. "From Primitive Accumulation to Regimes of Dispossession: Six Theses on India's Land Question." *Economic and Political Weekly* 50 (22): 146–57.

Lévi-Strauss, Claude. 1966. *The Savage Mind (La pensée sauvage).* London: Weidenfeld & Nicolson.

Linkenbach, Antje. 2007. *Forest Futures: Global Representations and Ground Realities in the Himalayas.* London: Seagull.

Livingston, Julie. 2012. *Improvising Medicine: An African Oncology Ward in an Emerging Cancer Epidemic.* Durham, NC: Duke University Press.

Locke, Piers. 2017. "Elephants as Persons, Affective Apprenticeship, and Fieldwork with Nonhuman Informants in Nepal." *HAU: Journal of Ethnographic Theory* 7 (1): 353–76.

Lutgendorf, Philip. 2007. *Hanuman's Tale: The Messages of a Divine Monkey.* Oxford: Oxford University Press, 2007.

Macintyre, Donald. 1889. *Hindu-Koh: Wanderings and Wild Sport on and beyond the Himalayas.* Edinburgh: W. Blackwood.

Maestripieri, Dario. 2007. *Macachiavellian Intelligence: How Rhesus Macaques and Humans Have Conquered the World.* Chicago: University of Chicago Press.

Malik, Aditya. 2016. *Tales of Justice and Rituals of Divine Embodiment: Oral Narrative from the Central Himalayas.* New York: Oxford University Press.

Malik, Iqbal. 1992. "Introduction." In *Primatology in India*, edited by M. H. Schwibbe and Iqbal Malik, 3–4. Primate Report, 34. Göttingen: DPZ.

Malik, Iqbal, and Rodney L. Johnson. 1994. "Commensal Rhesus in India: The Need and Cost of Translocation." *Revue d'Ecologie (Terre Vie)* 49: 233–43.

Malik, Iqbal, P. K. Seth, and Charles H. Southwick. 1984. "Population Growth of Free-Ranging Rhesus Monkeys at Tughlaqabad." *American Journal of Primatology* 7 (4): 311–21.

———. 1985. "Group Fission in Free-Ranging Rhesus Monkeys of Tughlaqabad, Northern India." *International Journal of Primatology* 6 (4): 411–22.

Massumi, Brian. 2002. *Parables for the Virtual: Movement, Affect, Sensation. Post-Contemporary Interventions*. Durham, NC: Duke University Press.

———. 2014. *What Animals Teach Us about Politics*. Durham, NC: Duke University Press, 2014.

Mathur, Nayanika. 2015. "'It's a Conspiracy Theory and Climate Change': Of Beastly Encounters and Cervine Disappearances in Himalayan India." *HAU: Journal of Ethnographic Theory* 5 (1): 87–111.

———. 2016. *Paper Tiger: Law, Bureaucracy and the Developmental State in Himalayan India*. Delhi: Cambridge University Press.

Mawdsley, Emma. 1998. "After Chipko: From Environment to Region in Uttaranchal." *Journal of Peasant Studies* 25 (4): 36–54.

———. 2002. "Redrawing the Body Politic: Federalism, Regionalism and the Creation of New States in India." *Commonwealth & Comparative Politics* 40 (3): 34–54.

Mazzarella, William. 2009. "Affect: What Is It Good For?" In *Enchantments of Modernity: Empire, Nation, Globalization*, edited by Saurabh Dube, 291–309. London: Routledge.

Mbembe, Achille. 2001. *On the Postcolony*. Berkeley: University of California Press.

McLane, John R. 1977. *Indian Nationalism and the Early Congress*. Princeton, NJ: Princeton University Press.

McLean, Stuart. 2016. "'Nature.' Theorising the Contemporary," *Cultural Anthropology* website, January 21. https://culanth.org/fieldsights/789-nature.

Messner, Reinhold. 2000. *My Quest for the Yeti: The World's Greatest Mountain Climber Confronts the Himalayas' Deepest Mystery*. New York: St. Martin's Press.

Mikhail, Alan. 2011. *Nature and Empire in Ottoman Egypt: An Environmental History*. Cambridge: Cambridge University Press.

Mines, Diane P., and Nicolas Yazgi. 2010. *Village Matters: Relocating Villages in the Contemporary Anthropology of India*. New Delhi: Oxford University Press.

Mittermaier, Amira. 2011. *Dreams That Matter: Egyptian Landscapes of the Imagination*. Berkeley: University of California Press.

Moffatt, Michael. 1979. *An Untouchable Community in South India: Structure and Consensus*. Princeton, NJ: Princeton University Press.

Mohanty, Talpade Chandra. 1988. "Under Western Eyes: Feminist Scholarship and Colonial Discourses." *Feminist Review* 30 (1): 61–88.

Morton, Timothy. 2011. "Here Comes Everything: The Promise of Object-Oriented Ontology." *Qui Parle* 19 (2): 163–90.

Müenster, Daniel. 2017. "Zero Budget Natural Farming and Bovine Entanglements in South India." *RCC Perspectives: Transformations in Environment and Society* 1: 25–32.

Muñoz, José Esteban. 2009. *Cruising Utopia: The Then and There of Queer Futurity; Sexual Cultures*. New York: New York University Press.

Nadasdy, Paul. 2007. "The Gift in the Animal: The Ontology of Hunting and Human-Animal Sociality." *American Ethnologist* 34 (1): 25–43.

Nading, Alex M. 2014. *Mosquito Trails: Ecology, Health, and the Politics of Entanglement.* Berkeley: University of California Press.

Narayan, Kirin. 1989. *Storytellers, Saints, and Scoundrels: Folk Narrative in Hindu Religious Teaching.* Philadelphia: University of Pennsylvania Press.

Narayan, Kirin, and Urmila Devi Sood. 1997. *Mondays on the Dark Night of the Moon: Himalayan Foothill Folktales.* New York: Oxford University Press.

Neumann, Roderick P. 1998. *Imposing Wilderness: Struggles over Livelihood and Nature Preservation in Africa.* Berkeley: University of California Press.

Neville, Melvin K. 1968. "Ecology and Activity of Himalayan Foothill Rhesus Monkeys (*Macaca mulatta*)." *Ecology* 49 (1): 110–23.

Nyong'o, Tavia. 2015. "Little Monsters: Race, Sovereignty, and Queer Inhumanism in Beasts of the Southern Wild." *GLQ: A Journal of Lesbian and Gay Studies* 21 (2/3): 249–72.

Ogra, Monica. 2009. "Attitudes toward Resolution of Human-Wildlife Conflict among Forest-Dependent Agriculturalists near Rajaji National Park, India." *Human Ecology* 37 (2): 161–77.

Ohnuki-Tierney, Emiko. 1974. "Another Look at the Ainu: A Preliminary Report." *Arctic Anthropology* 11: 189–95.

———. 1987. *The Monkey as Mirror: Symbolic Transformations in Japanese History and Ritual.* Princeton, NJ: Princeton University Press.

Ortner, Sherry B. 1973. "On Key Symbols." *American Anthropologist* 75(5): 1338–46.

Pachirat, Timothy. 2011. *Every Twelve Seconds: Industrialized Slaughter and the Politics of Sight.* New Haven, CT: Yale University Press.

Pande, Vasudha. 2015. *Making Kumaun Modern: Family and Custom c. 1815–1930.* Nehru Memorial Museum and Library Occasional Paper. New Delhi: Nehru Memorial Museum and Library.

Pandey, Gyanendra. 1983. "Rallying around the Cow: Sectarian Strife in the Bhojpur Region, c. 1888–1917." In *Subaltern Studies: Writings on South Asian History and Society*, vol. 2, edited by Ranajit Guha, 60–129. New Delhi: Oxford University Press.

Pandian, Anand. 2008. "Devoted to Development: Moral Progress, Ethical Work, and Divine Favor in South India." *Anthropological Theory* 8 (2): 159–79.

———. 2009. *Crooked Stalks: Cultivating Virtue in South India.* Durham, NC: Duke University Press.

Pandian, Anand S. 2001. "Predatory Care: The Imperial Hunt in Mughal and British India." *Journal of Historical Sociology* 14 (1): 79–107.

Pandian, M. S. S. 1995. "Gendered Negotiations: Hunting and Colonialism in the Late 19th Century Nilgiris." *Contributions to Indian Sociology* 29 (1/2): 239–63.

Parreñas, Juno Salazar. 2016. "The Materiality of Intimacy in Wildlife Rehabilitation: Rethinking Ethical Capitalism through Embodied Encounters with Animals in Southeast Asia." *Positions* 24 (1): 97–127.

Parreñas, Rhacel Salazar. 2001. *Servants of Globalization: Women, Migration and Domestic Work.* Stanford, CA: Stanford University Press.

Parreñas, Rheana "Juno" Salazar. 2012. "Producing Affect: Transnational Volunteerism in a Malaysian Orangutan Rehabilitation Center." *American Ethnologist* 39 (4): 673–87.

Parry, Jonathan P. 1994. *Death in Banaras.* Cambridge: Cambridge University Press.

———.2008. "The Sacrifices of Modernity in a Soviet-Built Steel Town in Central India." In *On the Margins of Religion*, edited by Frances Pine and João De Pina-Cabral, 233–62. New York: Berghahn Books.

Pathak, Shekhar. 1997. "State, Society and Natural Resources in Himalaya: Dynamics of

Change in Colonial and Post-Colonial Uttarakhand." *Economic and Political Weekly* 32 (17): 908–12.

Philo, Chris, and Chris Wilbert. 2000. A*nimal Spaces, Beastly Places: New Geographies of Human-Animal Relations*. Critical Geographies. London: Routledge.

Pinney, Christopher. 2004. *Photos of the Gods: The Printed Image and Political Struggle in India*. London: Reaktion.

Pinto, Sarah. 2006. "Globalizing Untouchability: Grief and the Politics of Depressing Speech." *Social Text* 24 (1 86): 81–102.

Plumwood, Val. 2008. "Tasteless: Towards a Food-Based Approach to Death." *Environmental Values* 17 (3): 323–30.

Pocock, David Francis. 1973. *Mind, Body, and Wealth: A Study of Belief and Practice in an Indian Village*. Totowa, NJ: Rowman and Littlefield.

Porter, Natalie. 2013. "Bird Flu Biopower: Strategies for Multispecies Coexistence in Việt Nam." *American Ethnologist* 40 (1): 132–48.

Povinelli, Elizabeth A. 2001. "Radical Worlds: The Anthropology of Incommensurability and Inconceivability." *Annual Review of Anthropology* 30 (1): 319–34.

———. 2006. *The Empire of Love: Toward a Theory of Intimacy, Genealogy, and Carnality*. Durham, NC: Duke University Press.

Radhakrishna, Sindhu, Michael A. Huffman, and Anindya Sinha, eds. 2012. *The Macaque Connection: Cooperation and Conflict between Humans and Macaques*. New York: Springer.

Raffles, Hugh. 2002. *In Amazonia: A Natural History*. Princeton, NJ: Princeton University Press.

———. 2010. *Insectopedia*. New York: Pantheon Books.

———. 2012. "Twenty-Five Years Is a Long Time." *Cultural Anthropology* 27 (3): 526–34.

Raheja, Gloria Goodwin. 1988. *The Poison in the Gift: Ritual, Prestation, and the Dominant Caste in a North Indian Village*. Chicago: University of Chicago Press.

Raheja, Gloria Goodwin, and Ann Grodzins Gold. 1994. *Listen to the Heron's Words: Reimagining Gender and Kinship in North India*. Berkeley: University of California Press.

Ramanujan, A. K. 1991a. "Three Hundred Rāmāyanas: Five Examples and Three Thoughts on Translation." In *Many Rāmāyanas: The Diversity of a Narrative Tradition in South Asia*, edited by Paula Richman, 22–49. Berkeley: University of California Press.

———. 1991b. "Toward a Counter-System: Women's Tales." In *Gender, Genre, and Power in South Asian Expressive Traditions*, edited by Arjun Appadurai, Frank J. Corom, and Margaret Ann Mills, 33–35. Philadelphia: University of Pennsylvania Press.

Ramanujan, A. K., Stuart H. Blackburn, and Alan Dundes. 1997. *A Flowering Tree and Other Oral Tales from India*. Berkeley: University of California Press.

Ramberg, Lucinda. 2014. *Given to the Goddess: South Indian Devadasis and the Sexuality of Religion*. Durham, NC: Duke University Press.

Rangan, Haripriya. 1997. "Property vs. Control: The State and Forest Management in the Indian Himalaya." *Development and Change* 28 (1): 71–94.

———. 2000. *Of Myths and Movements: Rewriting Chipko into Himalayan History*. London: Verso.

Rangarajan, Mahesh. 1998. "The Raj and the Natural World: The War against 'Dangerous Beasts' in Colonial India." *Studies in History* 14 (2): 265–99.

Rangarajan, Mahesh, and Ranthambhore Foundation. 2001. *India's Wildlife History: An Introduction*. Delhi: Permanent Black.

Rao, Anupama. 2009. *The Caste Question: Dalits and the Politics of Modern India*. Berkeley: University of California Press.

Richard, A. F., S. J. Goldstein, and R. E. Dewar. 1989. "Weed Macaques: The Evolutionary Implications of Macaque Feeding Ecology." *International Journal of Primatology* 10 (6): 569–94.

Ring, Laura A. 2006. *Zenana: Everyday Peace in a Karachi Apartment Building*. Bloomington:
Indiana University Press.

Ritvo, Harriet. 2004. "Animal Planet." *Environmental History* 9 (2): 204–20.

———. 2007. "On the Animal Turn." *Daedalus* 136 (4): 118–22.

Roychowdhury, Poulami. 2013. "The Delhi Gang Rape: The Making of International Causes."
Feminist Studies 39 (1): 282–92.

Rubin, Gayle. 1975. "The Traffic in Women: Notes on the 'Political Economy' of Sex." In *Toward
an Anthropology of Women*, edited by Rayna R. Reiter, 157–210: New York: Monthly Review
Press, 1975.

Ruel, Malcolm. (1982) 2002. "Christians as Believers." In *Religious Organization and Religious
Experience*, edited by J. Davis. London: Academic Press. Reprint in *A Reader in the Anthropology of Religion*, edited by Michael Lambek, 99–113, Malden, MA: Blackwell.

Rutherford, Danilyn. 2012. "Commentary: What Affect Produces." *American Ethnologist* 39 (4):
688–91.

Saberwal, Vasant, and Mahesh Rangarajan. 2009. *Battles over Nature: Science and the Politics of
Conservation*. Delhi: Permanent Black.

Saberwal, Vasant, Mahesh Rangarajan, and Ashish Kothari. 2001. *People, Parks, and Wildlife:
Towards Coexistence*. New Delhi: Orient Longman.

Saha, Jonathan. 2016. "Milk to Mandalay: Dairy Consumption, Animal History and the Political
Geography of Colonial Burma." *Journal of Historical Geography* 54: 1–12.

Sarkar, Tanika. 2001. *Hindu Wife, Hindu Nation: Community, Religion, and Cultural Nationalism*.
New Delhi: Permanent Black.

Sathyakumar, Sambandam. 2001. "Status and Management of Asiatic Black Bear and Himalayan
Brown Bear in India." *Ursus* 12: 21–29.

Sax, William Sturman. 2009. *God of Justice: Ritual Healing and Social Justice in the Central Himalayas*. New York: Oxford University Press.

Schneider, David Murray. 1984. *A Critique of the Study of Kinship*. Ann Arbor: University of
Michigan Press.

Selby, Martha Ann. 2011. *Tamil Love Poetry: The Five Hundred Short Poems of the Ainkurunuru,
an Early Third Century Anthology*. New York: Columbia University Press.

Sengupta, Asmita, Kim R. McConkey, and Sindhu Radhakrishna. 2015. "Primates, Provisioning
and Plants: Impacts of Human Cultural Behaviours on Primate Ecological Functions." *PLoS
ONE* 10 (11): e0140961.

Shah, Alpa. 2010. *In the Shadows of the State Indigenous Politics, Environmentalism, and Insur-
gency in Jharkhand, India*. Durham, NC: Duke University Press.

Shiva, Vandana. 1988. *Staying Alive: Women, Ecology, and Development*. London: Zed Books.

Shresth, Swati. 2009. "Sahibs and Shikar: Colonial Hunting and Wildlife in British India, 1800–
1935." PhD diss., Duke University.

Singh, Bhrigupati. 2011. "Agonistic Intimacy and Moral Aspiration in Popular Hinduism: A Study
in the Political Theology of the Neighbor." *American Ethnologist* 38 (3): 430–50.

———. 2015. *Poverty and the Quest for Life: Spiritual and Material Striving in Rural India*. Chicago: University of Chicago Press.

Singh, Bhrigupati, and Naisargi Dave. 2015. "On the Killing and Killability of Animals: Nonmoral
Thoughts for the Anthropology of Ethics." *Comparative Studies of South Asia, Africa and the
Middle East* 35 (2): 232–45.

Singh, S. D. 1968. "Social Interactions between the Rural and Urban Monkeys, Macaca Mulatta." *Primates* 9 (1): 69–74.

Sinha, Anindya. 2003. "A Beautiful Mind: Attribution and Intentionality in Wild Bonnet Macaques." *Current Science* 85 (7): 1021–30.

———. 2005. "Not in Their Genes: Phenotypic Flexibility, Behavioural Traditions and Cultural Evolution in Wild Bonnet Macaques." *Journal of Biosciences* 30 (1): 51–64.

Sivaramakrishnan, K. 1999. *Modern Forests: Statemaking and Environmental Change in Colonial Eastern India.* Stanford, CA: Stanford University Press.

———. 2011. "Thin Nationalism: Nature and Public Intellectualism in India." *Contributions to Indian Sociology* 45 (1): 85–111.

Sivaramakrishnan, Kalyanakrishnan, and Arun Agrawal, eds. 2003. *Regional Modernities: The Cultural Politics of Development in India.* Stanford, CA: Stanford University Press.

Skaria, Ajay. 1999. *Hybrid Histories: Forests, Frontiers, and Wildness in Western India.* Studies in Social Ecology and Environmental History. Delhi: Oxford University Press.

Smith, Brian K., and Wendy Doniger. 1989. "Sacrifice and Substitution: Ritual Mystification and Mythical Demystification." *Numen* 36 (2): 189–224.

Smith, W. Robertson. (1889) 1927. *Lectures on the Religion of the Semites: The Fundamental Institutions.* 3rd ed. New York: Macmillan.

Smuts, Barbara. 1999. "Reflections." In *The Lives of Animals,* by J. M. Coetzee. Princeton, NJ: Princeton University Press.

Solomon, Daniel Allen. 2013. "Menace and Management: Power in the Human-Monkey Social Worlds of Delhi and Shimla." PhD diss., University of California, Santa Cruz.

Southwick, Charles H., and M. Farooq Siddiqi. 1988. "Partial Recovery and a New Population Estimate of Rhesus Monkey Populations in India." *American Journal of Primatology* 16 (3): 187–97.

———. 1994. "Primate Commensalism: The Rhesus Monkey in India." *Revue d'Ecologie (Terre Vie)* 49: 223–31.

Stefan, Fiol. 2013. "Of Lack and Loss: Assessing Cultural and Musical Poverty in Uttarakhand." *Yearbook for Traditional Music* 45: 83–96.

Stoler, Ann Laura. 2013. *Imperial Debris: On Ruins and Ruination.* Durham, NC: Duke University Press.

Strathern, Marilyn. 2004. *Partial Connections.* Walnut Creek, CA: AltaMira.

TallBear, Kim. 2011. "Why Interspecies Thinking Needs Indigenous Standpoints." Theorizing the Contemporary, *Cultural Anthropology* website, April 24. https://culanth.org/fieldsights/260 -why-interspecies-thinking-needs-indigenous-standpoints.

Taneja, Anand. 2017. *Jinnealogy: Time, Islam, and Ecological Thought in the Medieval Ruins of Delhi.* Stanford, CA: Stanford University Press.

Taussig, Michael T. 1987. *Shamanism, Colonialism, and the Wild Man: A Study in Terror and Healing.* Chicago: University of Chicago Press.

Teas, Jane, Henry A. Feldman, Thomas L. Richie, Henry G. Taylor, and Charles H. Southwick. 1982. "Aggressive Behavior in the Free-Ranging Rhesus Monkeys of Kathmandu, Nepal." *Aggressive Behavior* 8 (1): 63–77.

Todd, Zoe. 2016. "An Indigenous Feminist's Take on the Ontological Turn: 'Ontology' Is Just Another Word for Colonialism." *Journal of Historical Sociology* 29 (1): 4–22.

Topsell, Edward, Conrad Gessner, and Thomas Moffett. (1658) 1967. *The History of Four-Footed Beasts and Serpents and Insects.* 3 vols. Reprint, New York: Da Capo.

Trawick, Margaret. 1988. "Spirits and Voices in Tamil Songs." *American Ethnologist* 15 (2): 193–215.

Trouillot, Michel Rolph. 1991. "Anthropology and the Savage Slot: The Poetics and Politics of Otherness." In *Recapturing Anthropology: Working in the Present*, edited by Richard G. Fox, 17–44. New Brunswick, NJ: Rutgers University Press.

Tsing, Anna. 2012. "Unruly Edges: Mushrooms as Companion Species: For Donna Haraway." *Environmental Humanities* 1 (1): 141–54.

———. 2015. *The Mushroom at the End of the World: On the Possibility of Life in Capitalist Ruins.* Princeton, NJ: Princeton University Press.

Tylor, Edward B. 1871. *Primitive Culture: Researches into the Development of Mythology, Philosophy, Religion, Art, and Custom.* London: J. Murray.

Van der Veer, Peter. 1994. *Religious Nationalism: Hindus and Muslims in India.* Berkeley: University of California Press.

Van Dooren, Thom, Eben Kirksey, and Ursula Münster. 2016. "Multispecies Studies: Cultivating Arts of Attentiveness." *Environmental Humanities* 8 (1): 1–23.

Van Dooren, Thom, and Deborah Bird Rose. 2012. "Storied-Places in a Multispecies City." *Humanimalia* 3 (2): 1–27.

Viveiros de Castro, Eduardo. 2004. "Perspectival Anthropology and the Method of Controlled Equivocation." *Tipití: Journal of the Society for the Anthropology of Lowland South America* 2 (1): 3–22.

———. 2012. "Immanence and Fear: Stranger-Events and Subjects in Amazonia." *HAU: Journal of Ethnographic Theory* 2 (1): 27–43.

———. 2015. *The Relative Native.* Translated by Julia Sauma, David Rogers, and Martin Holbraad. Chicago: Hau Books.

Wadley, Susan Snow. 1994. *Struggling with Destiny in Karimpur, 1925–1984.* Berkeley: University of California Press.

Weaver, Harlan. 2013. "Becoming in Kind: Race, Class, Gender, and Nation in Cultures of Dog Rescue and Dogfighting." *American Quarterly* 65 (3): 689–709.

———. 2015. "Pit Bull Promises: Inhuman Intimacies and Queer Kinships in an Animal Shelter." *GLQ: A Journal of Lesbian and Gay Studies* 21 (2/3): 343–63.

White, Sam. 2011. "From Globalized Pig Breeds to Capitalist Pigs: A Study in Animal Cultures and Evolutionary History." *Environmental History* 16 (1): 94–120.

Yang, Anand A. 1980. "Sacred Symbol and Sacred Space in Rural India: Community Mobilization in the 'Anti-Cow Killing' Riot of 1893." *Comparative Studies in Society and History* 22 (4): 576–96.

Zigon, Jarrett. 2014. "An Ethics of Dwelling and a Politics of World-Building: A Critical Response to Ordinary Ethics." *Journal of the Royal Anthropological Institute* 20 (4): 746–64.

Index

Page numbers in italics refer to illustrations.